D0076461

Elliott Sober is a leading philosopher of science and is a winner of the Lakatos Prize, the major award in the field. This new collection of essays will appeal to a readership that extends well beyond the frontiers of the philosophy of science. Sober shows how ideas in evolutionary biology bear in significant ways on traditional problems in philosophy of mind and language, epistemology, and metaphysics. Among the topics addressed are psychological egoism, solipsism, the interpretation of belief and utterance, empiricism, Ockham's razor, causality, essentialism, and scientific laws.

Sober's collection is informed by the twin beliefs that the contents of biological theories speak to problems that first saw the light of day as problems in philosophy, and that biological theories provide philosophically interesting examples of how the enterprise of knowledge proceeds. By demonstrating how a philosopher has come to understand philosophical problems by understanding scientific problems and thus how these two worlds of inquiry are interrelated, this volume is paradigmatic of the broader goals of this series.

The collection will prove invaluable to a wide range of philosophers, primarily those working in the philosophy of science, the philosophy of mind, and epistemology.

From a biological point of view

CAMBRIDGE STUDIES IN PHILOSOPHY AND BIOLOGY

General editor

Michael Ruse *University of Guelph*

Advisory board

Michael Donoghue *Harvard University*
Jonathan Hodge *University of Leeds*
Jane Maienschein *Arizona State University*
Jesus Mosterin *University of Barcelona*
Elliott Sober *University of Wisconsin*

This major new series will publish the very best work in the philosophy of biology. Nonsectarian in character, the series will extend across the broadest range of topics: evolutionary theory, population genetics, molecular biology, ecology, human biology, systematics, and more. A special welcome will be given to contributions treating significant advances in biological theory and practice, such as those emerging from the Human Genome Project. At the same time, due emphasis will be given to the historical context of the subject, and there will be an important place for projects that support philosophical claims with historical case studies.

Books in the series will be genuinely interdisciplinary, aimed at a broad cross section of philosophers and biologists, as well as interested scholars in related disciplines. They will include specialist monographs, collaborative volumes, and – in a few instances – selected papers by a single author.

Forthcoming

Alfred I. Tauber *The Immune Self: Theory or Metaphor?*
Peter Godfrey-Smith *Teleonomy and the Philosophy of Mind*

From a biological point of view

Essays in evolutionary philosophy

ELLIOTT SOBER
University of Wisconsin, Madison

Published by the Press Syndicate of the University of Cambridge
The Pitt Building, Trumpington Street, Cambridge CB2 1RP
40 West 20th Street, New York, NY 10011–4211, USA
10 Stamford Road, Oakleigh, Melbourne 3166, Australia

© Cambridge University Press 1994

First published 1994

Printed in the United States of America

Library of Congress Cataloging-in-Publication Data

Sober, Elliott.

From a biological point of view : essays in evolutionary philosophy / Elliott Sober.

p. cm. – (Cambridge studies in philosophy and biology)

Includes index.

ISBN 0-521-47184-2 (hc). – ISBN 0-521-47753-0 (pb)

1. Evolution. 2. Biology – Philosophy. 3. Philosophy of mind. 4. Ethics, Evolutionary. 5. Knowledge, Theory of. 6. Empiricism. 7. Essence (Philosophy) 8. Time. I. Title. II. Series.
B818.S64 1994
146′.7 – dc20 94-15745
CIP

A catalog record for this book is available from the British Library.

ISBN 0–521–47184–2 hardback
ISBN 0–521–47753–0 paperback

For
Paul Korshin, Paul Fitzgerald, and
Mary Hesse

Contents

Contents

Acknowledgments

Essays 1, 3, and 4 appear here for the first time. The other essays are identical, minor revisions excepted, with essays that have appeared or will soon appear in the publications listed below.

2. "Why Not Solipsism?" *Philosophy and Phenomenological Research,* forthcoming. Reprinted by permission of *Philosophy and Phenomenological Research.*
5. "Prospects for an Evolutionary Ethics." In L. Pojman (ed.), *Ethical Theory,* Wadsworth, 1994. Reprinted by permission of the editor.
6. "Contrastive Empiricism." In Wade Savage (ed.), *Minnesota Studies in the Philosophy of Science (vol. 14): Scientific Theories* (University of Minnesota Press, 1990), pp. 392–412. Reprinted by permission of University of Minnesota Press.
7. "Let's Razor Ockham's Razor." In D. Knowles (ed.), *Explanation and Its Limits,* Royal Institute of Philosophy Supplementary Volume 27 (Cambridge University Press, 1990), pp. 73–94. Reprinted by permission of Cambridge University Press.
8. "The Principle of the Common Cause." In J. Fetzer (ed.), *Probability and Causation: Essays in Honor of Wesley Salmon* (Dordrecht: Reidel, 1988), pp. 211–28. Reprinted by permission of Kluwer Academic Publishers.
9. "Explanatory Presupposition." *Australasian Journal of Philosophy, 64* (1986), 143–9. Reprinted by permission of the *Australasian Journal of Philosophy.*
10. "Apportioning Causal Responsibility." *Journal of Philosophy 85,* (1988), 303–18. Reprinted by permission of the *Journal of Philosophy.*
11. "Evolution, Population Thinking, and Essentialism." *Philosophy of*

Science, 47, 3 (1980), 350–83. Reprinted by permission of the Philosophy of Science Association.

12. "Temporally Oriented Laws." *Synthese 94* (1993), 171–89. Reprinted by permission of Kluwer Academic Publishers.

Introduction

Philosophers have tried to learn from science in two quite different ways. First, the contents of particular scientific theories have thrown light on philosophical problems. For example, philosophers interested in the nature of space and time have found much of interest in relativity theory. As sometimes happens in the history of thought, a problem that begins its life as a problem in philosophy later turns out to receive illumination from a body of scientific results. Philosophy then has to catch up with the news from outside.

A second avenue of influence develops when philosophers contemplate the character of science itself. Here scientific theories serve as examples. Relativity theory says nothing about what scientific knowledge is; its subject is space, time, and motion, not the nature of inquiry. Still, the hypotheses of relativity theory have been an important point of reference for philosophers seeking to understand how the human mind is able to grasp a reality outside itself.

I've called this collection of papers *From a Biological Point of View* because it is biology – particularly the theory of evolution – that has performed, in my own work, the two functions just described. I believe that the contents of biological theories speak to problems that first saw the light of day as problems in philosophy. And I have continually been struck by the fact that biological theories provide philosophically interesting examples of how the enterprise of knowledge proceeds.

The title I've chosen for this volume is an homage to, and is intended to contrast with, W. V. Quine's collection *From a Logical Point of View* (Harvard University Press, 1953). In his work in metaphysics, epistemology, and other areas, Quine constantly reached for logical ideas as tools of clarification. Although Quine developed his views in opposition to logical empiricism and positivism, he shared with those he opposed a practical commitment to the centrality of logic as a tool of philosophy. I think it is fair to say that in the forty years since Quine's book appeared, logic and philosophy of science (to say nothing of logic and philosophy as a whole) have drifted further apart. Logic has edged closer to mathematics and computer science. And philosophers have become more skeptical that logic provides general techniques for solving philosophical problems. No new overarching paradigm now commands wide acceptance, not do I recommend my own approach as a panacea. Perhaps the problems we call "philosophical" are too diverse for any single method to achieve what logic was once thought to be able to do.

A work of philosophy should attempt to solve a philosophical problem. A work of *meta*philosophy, which is what this Introduction purports to be, discusses the character of philosophy itself. Metaphilosophy is a pursuit that inspires me with feelings of suspicion. Philosophical programs, I believe, should be judged by their fruits; they can't be justified in advance. What is more, philosophers often misrepresent the real nature of their methods when they self-consciously identify themselves with this or that program. With the demise of positivism and the rise of philosophical "naturalism," it has become fashionable to describe philosophy as a project that is continuous with the sciences. The idea that philosophers go in for conceptual analysis whereas scientists do empirical work is supposed to be an untenable dualism that sophisticates are expected to recognize as naive. Yet, in spite of this change in the metaphilosophical landscape, much of philosophy proceeds as it did before. It now is unfashionable to say that one is giving a conceptual analysis of the meaning of terms such as "justice," "knowledge," or "freedom"; instead, it is more acceptable to say that one is trying to describe what justice, knowledge, or freedom *is.* It is worth pondering how much of this shift is substantive, and how much is just window-dressing.

These cranky remarks should not be taken to suggest that I subscribe to some rigid separation of conceptual clarification and empirical insight. I love blurred boundaries as much as the next person. However, I do think that in *practice,* if not in *theory,* philosophy of science (and philosophy as a whole) largely remains a subject driven by its own problems and methods. It is no accident that we philosophers do not do experiments and do not go into the field to make natural observations. If metaphilosophy is what we want, we need a theory of philosophy that is true to the actual practice of that subject.

It should be no embarrassment to philosophy that its conundrums often fail to engage the attention of working scientists. If philosophy is a discipline in its own right, this is to be expected. Nor should a misplaced scientism lead us to think that philosophy is barred from making normative claims by way of criticizing what scientists do. I once heard a justly respected philosopher say that naturalism means that when philosophy and science conflict, it is almost always the philosopher who is wrong. This is correct, of course, when the philosopher is poorly informed and the scientist has his or her head screwed on straight. But we should not forget that scientists are mortals, and philosophers can learn from experience.

Even if philosophy is a discipline that is driven by its own evolving set of problems, this does not mean that scientific ideas are irrelevant to the solution of philosophical problems. What has happened in the past is doubtless happening right now. Science encroaches little by little on what philosophy views as its own terrain. Given the success of quantum mechanics as an empirical theory, we may be puzzled as to how earlier generations of philosophers could have thought that determinism is *a priori* true. Yet the idea that science right *now* may provide evidence that decides a problem that we *now* regard as philosophical will strike many philosophers as a quixotic dream.

The idea that philosophy is an *a priori* discipline and the idea that it is simply a part of science are both wrong. Philosophy is not a unity; different philosophical problems are structured differently. Nor can one tell in advance how one philosophical problem is related to others, and to matters that arise in other arenas of thought. We should relish the fact that philosophy can be *surprising.* Understanding the nature of a problem is not something we do in advance of trying to solve it.

When philosophy and science are brought into contact, the searchlight can be aimed from philosophy to science, from science to philosophy, or in both directions at once. An essay that is structured in the first way will address a problem that is of interest to working scientists; it will attempt to provide that scientific problem with philosophical illumination. If the essay succeeds, it will be of interest primarily to the scientists themselves. An essay of the second sort will identify a traditionally philosophical problem and will try to show how scientific ideas can be of use in its solution. Here the intended audience will be philosophers – even philosophers who have no antecedent reason to care about the scientific ideas addressed.

The essays in this volume are mainly of type two. They were written for philosophers, not for scientists (though I hope that philosophically inclined scientists may find some of the issues of interest). Although I believe that type one philosophy of science is important, I feel that my best efforts in that line of work are to be found in my books *The Nature of Selection, Reconstructing the Past,* and *Philosophy of Biology.* In assembling the present collection of essays, I have tried to follow the advice of Ockham's razor by not reprinting essays that I felt would be redundant.

The subjects explored in these essays are diverse. The first four address different features of the human mind that have long interested philosophers. Egoism, solipsism, innate beliefs, and the status of truth-telling and true belief in communication involve problems that are familiar to the philosopher of mind, the epistemologist, and the philosopher of language. What is somewhat less standard is the idea that these issues may be approached from an evolutionary point of view. Rather than ask what behavioral and introspective evidence there is for and against psychological egoism, I explore the question of whether evolution can be expected to have made us psychological egoists. And rather than ask what rational justification people can provide for their belief that a physical world exists external to their states of consciousness, I ask what adaptive advantage organisms obtain by thinking in nonsolipsistic terms. These evolutionary questions do not replace the traditional problems that prompt them. But I believe that they do have considerable philosophical interest in their own right.

Whereas the first four essays defend the bearing of biology on the philosophical problem at hand, the tone of Essay 5 is much more skeptical. Here I consider what evolution may be able to teach us about ethics. The fundamental distinction that needs to be drawn here is that between explaining why we have the ethical thoughts and feelings we do and deciding what status our ethical convictions possess. With respect to the first task, I emphasize the importance of thinking about *patterns of variation* as the fundamental *explanandum.* What needs to be explained is why individuals are similar and different in their ethical beliefs. This emphasis on variation as the proper object of explanation recurs in Essays 10 and 11.

The other main question addressed in Essay 5 concerns whether evolutionary considerations can show us whether our ethical beliefs are true. This is fundamentally an epistemological question – can the evolutionary origins of a belief undermine the claim that it is correct? Can they show, as ethical subjectivism maintains, that no ethical statements are ever true? Here we need to consider the status of *genetic arguments* and the significance of an influential argument that appeals to Ockham's razor – the principle of parsimony – in defense of ethical subjectivism.

Essay 6 addresses the dispute between scientific realists and empiricists. Realists see the discovery of true theories as the proper goal of science; empiricists think that science aims only at the discovery of predictively successful theories. Both philosophies embed particular views about how scientific inference works. In this essay, I propose a synthesis of realism and empiricism that I call *contrastive empiricism.* This position, I believe, retains what is plausible in each traditional doctrine, while avoiding the excesses of each.

Essays 7 and 8 elaborate and implement the epistemology laid out in Essay 6. The idea that we should use the simplicity of a theory to help decide whether the theory is plausible has always been a problem for empiricism. For Ockham's razor seems to say that plausibility is to be judged, in part, on *non*empirical grounds. The same problem arises in connection with the idea that correlated events should be explained by postulating a cause that they share. The challenge to empiricism is to explain the merits of these principles without departing from a reasonable empiricist epistemology.

Although Essays 6, 7, and 8 lay out the fundamentals of the epistemology I wish to defend, they leave considerable work undone. I believe that the essay on the curve-fitting problem that Malcolm Forster and I recently finished lends further credence to the view of simplicity and parsimony that I want to defend (Forster and Sober, 1994). And I have tried to be more precise about the concept of observation, and about the epistemological significance of the distinction between theory and observation, in Sober (1993a). In addition, I explore the bearing of contrastive empiricism on the use of indispensability arguments in philosophy of mathematics in Sober (1993b).

If science is the activity of putting questions to nature, then an important kind of scientific question formulates a request for explanation. What presuppositions do those why-questions have? In Essay 9, I argue that the explanatory problems we pose about nature embody particular assumptions about causal structure. The idea that different events should be traced back to a common cause is not just the inferential maxim discussed in Essay 7; it also is a guide to how we organize what we already know when we try to generate explanations.

The nature of causality – what it means for one event to cause another – is a perennial philosophical problem. A related, though distinct, problem concerns the nature of causal importance, the subject of

Essay 10. If an event has several causes, what does it mean to say that one of them is "more important" than another? This question repeatedly arises in biology and the social sciences, and the sciences have well-defined methods for handling them. Here is a metaphysical problem that the practice of biology informs.

Essentialism provides a view of science that many philosophers now associate with scientific realism. Just as science aims to discover unobservable causal mechanisms that account for observable phenomena, so it attempts to discover what the natural kinds are to which individuals entities belong. For a long time, the two standard examples of this essentialist interpretation of the activity of science have been the periodic table of elements in chemistry and the identification of biological species. In Essay 11, I argue that evolutionary theory undermines essentialism as a *global* philosophy of science, and does so for subtle reasons that have not been widely appreciated.

The last essay in this collection is on a problem in the philosophy of time. A law of nature can describe how a system in a given state will probably evolve or it can describe how a system in a given state probably did evolve. I call these two types of law *forward-directed* and *backward-directed,* respectively. It may seem at first glance that laws of nature can be of both sorts. However, I argue in Essay 12 that this appearance is misleading; when the concept of law is made precise in a certain way, one can show that systems that can be expected to evolve cannot have both sorts of laws. I also float a proposal concerning why science seems to opt for forward-directed more than backward-directed laws.

Even though the biological content of some of the essays I've described is obvious, others may seem to have only a scant connection with what might be termed "a biological point of view." After all, what could the principle of the common cause or a problem about time asymmetry have to do with the theory of evolution? Although a fuller answer to this question is contained in the essays themselves, I'll provide some brief comments here.

Evolutionists implicitly use the principle of the common cause when they argue that similarities between species should be explained by the hypothesis that the species have a common ancestor from which the similarities were inherited as homologies. It was this phylogenetic context that encouraged me to think about the justification of the principle of the common cause from a probabilistic point of view.

A similar biological background guided my thinking about the issue of temporal asymmetry discussed in Essay 12. The rules of Mendelian inheritance provide conditional probabilities that are forward-directed; they describe the probability that an offspring will have a particular genotype, given that the parents have this or that genotype. Those conditional probabilities do not change with time; even when the frequencies of the *A* and *a* genes evolve, the probability that an offspring will be *Aa,* given that its parents are both *Aa,* remains the same.

As natural and familiar as the Mendelian rules are, it is much harder to think about backward-directed probabilities. Why aren't the laws of inheritance stated by specifying the probability that both parents are heterozygotes, given that an offspring is a heterozygote? If such "laws" could be formulated, would they remain true as the population evolves? This simple biological example led me to consider a quite general problem about the character of scientific laws.

I would be lying if I described this collection of essays as a systematic and linear development of a single thesis. What unity the essays possess stems from the fact that I have tried to take seriously the content and the practice of evolutionary biology. The philosophical claims I defend are various; they do not stand or fall together. However, I would like to think that this variety, far from being a defect, reflects favorably on what a biological point of view in philosophy has to offer.

REFERENCES

Forster, M., and Sober, E. (1994). How to Tell When Simpler, More Unified, or Less *ad Hoc* Hypotheses Will Provide More Accurate Predictions. *British Journal for the Philosophy of Science, 45*: 136.

Sober, E. (1993a). "Epistemology for Empiricists." In H. Wettstein (ed.), *Midwest Studies in Philosophy,* volume 18, University of Notre Dame Press, pp. 39–61.

Sober, E. (1993b). Mathematics and Indispensability. *Philosophical Review, 102:* 35–57.

1

Did evolution make us psychological egoists?

1. TWO CONCEPTS

The concept of altruism has led a double life. In ordinary discourse, as well as in psychology and the social sciences, *altruism* refers to behaviors that are produced because people have certain sorts of motives. In evolutionary biology, on the other hand, the concept is applied to behaviors that enhance the fitness of others at expense to self.

A behavior can be altruistic in the evolutionary sense without being an example of psychological altruism. A plant that leeches insecticide into the soil may be an altruist, if the insecticide benefits its neighbors and imposes an energetic cost on the producer. In saying this, I am not attributing a mind to the plant. Evolutionary altruism has to do with the fitness consequences of the behavior, not with the mechanisms inside the plant (mental or otherwise) that cause the plant to behave as it does.

Symmetrically, a behavior can be altruistic in the psychological sense without being an example of evolutionary altruism. If I give you a volume of Beethoven piano sonatas (or a package of contraceptives) out of the goodness of my heart, my behavior may be psychologically altruistic. However, the gift giving will not be an example of evolutionary altruism, if the present fails to augment your prospects for survival and reproductive success.

Although the concepts are different, they have a few things in common. Both point to causal explanations of the behaviors so labeled. If I say that a behavior is an example of psychological altruism, I am making a claim about the motives that produced the behavior. If I say

I am grateful to Daniel Batson, Don Moskowitz, and the members of the philosophy departments at University of Pittsburgh, Rutgers University (New Brunswick), and at University of Wisconsin (Madison) for useful discussion.

that a behavior is an example of evolutionary altruism, I am suggesting a certain sort of explanation, which I'll describe in the next section, for why the behavior evolved.[1]

The second common feature of the psychological and the evolutionary concepts is that both have been controversial and even unpopular in much scientific investigation. Psychological egoism, which claims that all of our ultimate motives are selfish, has viewed psychological altruism as a comforting illusion. Egoism has been the dominant position in all major schools of twentieth-century psychology (Batson 1991). And within evolutionary biology, the theory of the selfish gene has been hostile to the idea that evolution produces behaviors that help the group at the expense of the individual (Williams 1966; Dawkins 1976).

My goal in this paper is to clarify these concepts and to further discuss why they are logically independent of each other. Then, having separated them, I will attempt to bring them back into contact. I will explore the question of whether there are evolutionary considerations that help us decide whether we are ever psychological altruists.

2. EVOLUTIONARY ALTRUISM

Altruism has been an important subject for evolutionary theorizing ever since Darwin. I will not describe the history of how this subject has developed, nor will I discuss intricacies that are internal to various theories of current interest.[2] My modest goal in this section is to describe with more care what altruism and selfishness mean in an evolutionary context and to show how each trait is connected with its own picture of how natural selection has operated.

For the most part, Darwin viewed natural selection as a process in which organisms within the same breeding population compete with each other to survive and reproduce. His picture of competition was not the lion versus the lamb, but lions competing with lions and lambs with lambs. In this process, the traits that evolve are the ones that benefit the individual organism. Although Darwinism is sometimes described by saying that characteristics evolve "for the good of the species," this is a major distortion of how Darwin *usually* thought about natural selection.

Usually, but not *always.* There were a small number of occasions on which Darwin took seriously the idea that natural selection involves competition among objects other than individual organisms. One of the clearest expressions of this alternative occurs in his discussion of human morality in *The Descent of Man.* Here is Darwin's statement of the problem:

> It is extremely doubtful whether the offspring of the more sympathetic and benevolent parents, or of those which were the most faithful to their comrades, would be reared in greater number than the children of selfish and treacherous parents of the same tribe. He who was ready to sacrifice his life, as many a savage has been, rather than betray his comrades, would often leave no offspring to inherit his noble nature. The bravest men, who were always willing to come to the front in war, and who freely risked their lives for others would on average perish in larger numbers than other men. (Darwin 1871, p. 163)

Darwin's point is that if we consider a single tribe that contains both altruistic and selfish individuals, altruists will do *worse* than selfish individuals. If natural selection is the main cause of evolutionary change (as Darwin thought), and if natural selection causes fitter traits to increase in frequency and less fit traits to decline, why hasn't altruism altogether disappeared from human conduct? Here is the answer that Darwin suggests:

> It must not be forgotten that although a high standard of morality gives but a slight or no advantage to each individual man and his children over the other men of the same tribe, yet that an advancement in the standard of morality and an increase in the number of well-endowed men will certainly give an immense advantage to one tribe over another. (Darwin 1871, p. 166).

Although altruistic *individuals* do worse than selfish *individuals* in the same tribe, altruistic *groups* do better than selfish *groups.* Here Darwin was imagining a process of group selection, in which groups compete against each other. This picture of the process of natural selection differs markedly from his more customary formulation, in which organisms within a single population engage in a struggle for existence.

To make sense of the idea of evolutionary altruism, and of the process of group selection that is associated with it, one must be able to think simultaneously about the fitnesses of organisms and the fitnesses of groups of organisms. How are these two levels related? And since altruism is a behavior produced by an individual organism, how do these two kinds of fitness make it possible for altruism to evolve when there is group selection?

Figure 1.1 depicts some of the main conceptual ingredients. It shows that the fitness of an individual depends on two factors. Whether the

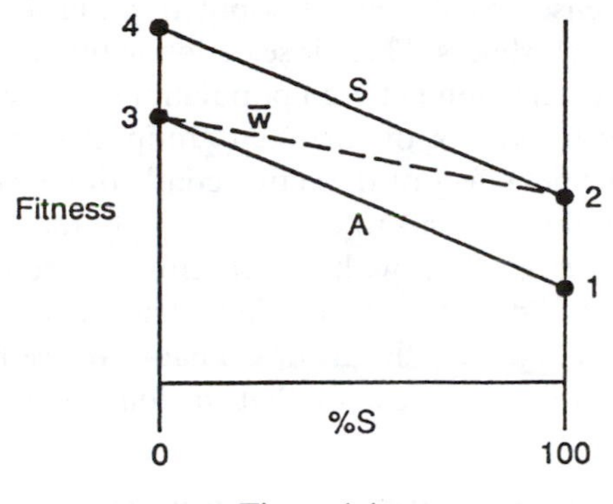

Figure 1.1

organism is altruistic (A) or selfish (S) influences its fitness; in addition, an individual's fitness also is influenced by the frequency of altruism in the group in which the individual lives.

Besides describing the fitnesses of the two traits, the figure also represents a third quantity, labeled $\bar{w}$ ("w-bar"). This is the average fitness of the organisms in the group. It is a plausible measure of what is meant by *the fitness of the group.* Interpreted in this way, the figure represents the idea that groups of altruists are fitter than groups of selfish individuals.

This figure expresses two ideas that are crucial to the definition of evolutionary altruism: (i) within each group, selfish individuals are fitter than altruists; (ii) groups of altruists are fitter than groups of selfish individuals. The question we now need to consider is how these two facts combine to determine whether altruism will evolve. We appear to have here a "conflict of interest" between what is good for the individual organism (selfishness) and what is good for the group (altruism). Which of these influences will be stronger?

If natural selection occurs within the confines of a single breeding population, then altruism cannot evolve. Even if the population somehow manages to be 100 percent altruistic, sooner or later a mutant or migrant selfish individual will appear. That selfish individual will be more reproductively successful than the average altruist, so selfishness will increase in frequency. In the next generation, selfishness still will be the fitter trait, and so the process eventually will carry the population all the way to 100 percent selfishness. This is the process that Dawkins (1976) called "subversion from within."

Matters change if we consider a more complicated scenario. We need to consider an ensemble of several populations, each with its own mix of altruism and selfishness. There is selection within each of these populations as well as selection between populations. The within-group process favors selfishness, but the between-group process favors altruism. What will be the net effect of these two conflicting processes?

The criterion for altruism to evolve is simply that altruism must be fitter than selfishness. True, within each group, altruism is less fit than selfishness. But this does not ensure that altruism is less fit than selfishness when we average over the groups. That is, to see how it is possible for altruism to evolve, we must see that the following argument is fallacious:

> Within each group, altruists have a lower fitness than selfish individuals.

> Hence, altruists have a lower fitness than selfish individuals in the ensemble of groups.

The premiss is part of the definition of evolutionary altruism. However, the conclusion does not follow.

To explain why this argument is fallacious, I will consider a simple example. Suppose that there are two groups. The first is 1 percent selfish and 99 percent altruistic; the second is 99 percent selfish and 1 percent altruistic. Let there be 100 individuals in each group. The fitnesses of the two traits, both within each group and averaging over the two groups, can be extracted from Figure 1.1 as follows:

Group 1	*Group 2*	*Global ensemble*
1(S): 4	99(S): 2	100(S): 2
99(A): 3	1(A): 1	100(A): 3

Notice that within each group, altruists are less fit than selfish individuals ($3 < 4$ and $1 < 2$). However, the global average is such that altruists are fitter than selfish individuals ($3 > 2$).

This decoupling of what is true within each group and what is true in the ensemble of groups is puzzling. It seems even more paradoxical if we consider what these numbers imply about the frequencies of the two traits in the next generation. What will happen is that altruism will decline in frequency within each group, but will increase in frequency in the two-group ensemble. The evolution of altruism requires that the fitnesses of the traits fit a pattern that statisticians call Simpson's Paradox (Sober 1984, 1988b).

Although altruism increases in frequency during the one generation time slice I have just described, it will not do so in the long term, if the two populations remain intact. Subversion from within will eliminate altruism from each subpopulation. And if altruism is absent in each part, it must be absent from the whole.

What is required for altruism to evolve by group selection is that groups go extinct and found colonies at different rates. In addition, these extinction and colonization events must occur often enough to offset the process that occurs within each group wherein selfishness replaces altruism. When these various assumptions are satisfied, an altruistic trait can evolve despite the fact that it is disadvantageous to the organisms possessing it.

The ideas just outlined do not show that altruism has evolved, either in our species or in others. Rather, I have simply sketched the biological assumptions that must be true if group selection is to have this result. It is a matter of continuing controversy in evolutionary biology how often this type of selection process actually occurs.

3. PSYCHOLOGICAL ALTRUISM

Let us turn now from the evolutionary issue to the psychological one. Is psychological altruism ever a part of human motivation or are our motives thoroughly egoistic? Before this question can be answered, I must clarify what the two psychological concepts mean.

It is quite clear that we sometimes help others. It also is clear that we sometimes want to do this. Consider the example of parental care. Human beings take care of their children (very often, though, unfortunately, not always). Moreover, this is something that parents want to do; the helping behavior stems from a desire to help.

It does not follow from the fact that we help our children, nor from the fact that we want to provide such help, that we are altruistically motivated. To tell whether parental care is an example of psychological altruism, we must ask *why* we help our children and *why* we want to provide this help.

The thesis of psychological egoism maintains that parents want to take care of their children only because parental care provides some sort of benefit to the parents. For example, it might be argued that parents who take care of their children experience various pleasures and avoid feeling guilty. For them, the welfare of their children is not an end in itself, but is merely a means to some more ultimate selfish goal.[3] The contrary position, which says that the behavior in question is at least sometimes altruistic, claims that parents have an irreducible interest in having their children do better rather than worse. The welfare of one's children is an end in itself, not merely a means to some selfish goal.

To clarify the difference between these two theories about human motivation, we need to understand what it means to say that *we want X only because X is a means to attaining Y.* Egoism affirms, and altruism denies, a claim of this form. I'll need to distinguish what I call the *self-directed* and the *other-directed* preferences an agent might have in some situation. A self-directed preference describes the situation of the agent, but not the situation of anyone else. An example would be my preferring that I have more money rather than less. An other-directed preference describes the situation of someone else, but not of the agent. An example would be my preferring that you have more money rather than less. In addition to these two "pure" sorts of preference, there are "mixed" preferences as well, which describe both the situation of self and the situation of other. An example would be my desire that you and I have the same income. Mixed preferences, though ubiquitous, can be left to one side.

I will suppose that agents decide which action to perform in a situation by seeing which of the available actions (by their lights) maximizes the satisfaction of their preferences.[4] For example, suppose you are contemplating whether to send a check for $25 to a charity. This action, you believe, will benefit malnourished children. It will cost you a modest amount of money. And there are the psychic consequences as well; you will think well of yourself and avoid feeling guilty. Suppose you send the check to the charity. The action you performed was the upshot of the various preferences you had. Perhaps you preferred that you feel good about yourself rather than bad; perhaps you also preferred that the children be better off rather than worse. For simplicity, I'll ignore additional preferences you may have had (including the one about money), and focus on just these two possibilities. Notice that one of them is self-directed and the other is other-directed.

There are four relationships that might obtain among these two possible preferences. They are depicted in Table 1.1. The numbers indicate a preference order; their absolute values have no significance.

Individuals with the preference structure I call *extreme egoism* care only about themselves; the welfare of others does not matter to them at all. Symmetrically, people with the preference structure I call *extreme altruism* care only about others, not about themselves. Each of these

Table 1.1

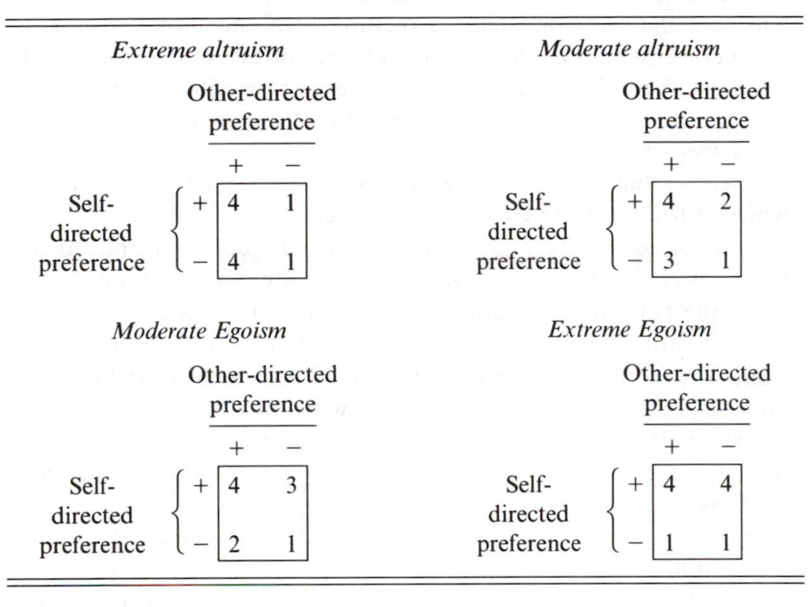

Extreme altruism

		Other-directed preference +	Other-directed preference −
Self-directed preference	+	4	1
	−	4	1

Moderate altruism

		Other-directed preference +	Other-directed preference −
Self-directed preference	+	4	2
	−	3	1

Moderate Egoism

		Other-directed preference +	Other-directed preference −
Self-directed preference	+	4	3
	−	2	1

Extreme Egoism

		Other-directed preference +	Other-directed preference −
Self-directed preference	+	4	4
	−	1	1

structures says that people are motivated by a single kind of preference. Note that when people face choices in which self-interest and the welfare of others *coincide,* they will perform the same action, regardless of whether they are extreme egoists or extreme altruists.[5] When the choice is between upper-left and lower-right, both extreme altruists and extreme egoists select the former (since $4 > 1$).

The other two structures depicted in Table 1.1 describe people who are motivated by two sorts of preference. *Moderate egoists* and *moderate altruists,* as I call them, care both about themselves and about others. However, they differ when they face a choice in which self-interest and the welfare of others *conflict.* When the choice is between upper-right and lower-left, a moderate egoist will place self-interest ahead of the welfare of others. A moderate altruist, in the same situation, will sacrifice self-interest for the sake of the other.

Moderate and extreme egoists are not willing to sacrifice self for the sake of another. Moderate and extreme altruists are willing to do so. The difference between altruism and egoism, when defined in this way, does not concern whether people *have* other-directed preferences, but whether self or other *matters more* in the choice situation at hand (Sober 1989).[6]

In describing these four arrangements, I do not assume that a person will have the same preference structure from one choice situation to another. Presumably, a person might be an altruist in one situation but an egoist in another. Even if we are prepared to sacrifice our own happiness for the sake of our children, few of us are prepared to die to make a stranger smile.

I believe that much of human behavior involves choice situations in which self-interest and the welfare of others coincide. We decide to give the money to charity, and this action benefits both self and others. I hope it is clear that behavior in such choice situations is entirely ambiguous; the behavior does not tell us what our motives are. All four of the motivational structures just described prefer upper-left over lower-right. The fact that we give money to charity does not show that we are altruists; and the fact that we feel good when we donate to charity does not show that we are egoists. Only when there is *conflict* between self-interest and the welfare of others does behavior reveal whether we are altruists or egoists.

According to this way of seeing things, psychological egoism and altruism are behavioral dispositions. These dispositions should not be confused with the behaviors they sometimes occasion. Altruism is not the same as helping. In addition, the dispositions thus defined should not be confused with the individual motives that may underlie them. Altruists are people disposed to sacrifice self-interest for the sake of others, if the two conflict. Yet it is perfectly possible for altruists to care about themselves; they may have irreducibly self-directed preferences. Symmetrically, egoists are people who are unwilling to sacrifice self-interest for the sake of others, if those should come into conflict. Yet it is perfectly possible for egoists to care about others; they may have irreducibly other-directed preferences. A consequence is that we should not conflate the following two questions:

(i) Do people ever have preferences concerning the welfare of others in which the welfare of others is not a means for advancing self-interest?

(ii) Are people ever willing to sacrifice self-interest for the sake of others when these conflict?

An affirmative answer to (ii) entails an affirmative answer to (i), but not conversely. If people are willing to sacrifice self-interest, not only do they care about others; in addition, they sometimes give more weight to the welfare of others than they do to self-interest.

It follows that we must keep separate the following two egoistic theses:

(E1) People never have irreducible preferences about the welfare of others.

(E2) People never are willing to sacrifice their own interests for the sake of others when these come into conflict.

(E1) entails (E2), but not conversely. (E1) rules out three of the preference structures I have described; (E2) rules out only two.

Although I presented the four preference structures in order to clarify what (E2) asserts, the same ideas can be used to explain what the word "irreducible" means in (E1). Altruists have irreducibly other-directed preferences; the same is true of moderate egoists. Extreme egoists are the only people for whom the welfare of others is a mere instrument for promoting the satisfaction of self-directed preferences.[7] All four of these characters will donate the $25 to charity, thereby causing both self and other to benefit. What is distinctive about extreme egoists is that they help *only because* they want to feel good about themselves.

4. PROXIMATE AND ULTIMATE EXPLANATION

Whenever natural selection causes a behavior to evolve, it must equip the organism with a mechanism that triggers the behavior in the appropriate circumstances (Mayr 1961). For example, if parental care is to evolve, the organism needs to have a device that causes it to dole out care to some organisms rather than to others.

For the organism to produce the selected behavior, it must possess both a *detector* and an *effector* (Williams 1966). For example, an organism may provide care to its offspring if it is programmed to drop food into gaping mouths. This will be a feasible solution to the design problem, if the parent can tell when gaping mouths are present and if the gaping mouths it sees are the mouths of its offspring.[8] If selection favors performing behavior *B* when and only when circumstance *C* obtains, the organism must have a detector of *C* and a way of producing *B* when *C* obtains.

Most organisms manage to evolve behavior without having minds as proximate mechanisms. Bacteria swim toward nutrients, but they do not form beliefs about where the nutrients are nor do they have desires concerning what is good to eat. We human beings, on the other hand, are guided in much of what we do by the beliefs and desires we have. It follows that if natural selection has caused human beings to behave in certain ways, it is plausible to suspect that natural selection has done this by equipping us with beliefs and desires that produce the selected behaviors in the appropriate circumstances.

If natural selection has worked strictly by way of individual selection, then it will have produced behaviors that are evolutionarily selfish. On the other hand, if group selection has played a substantial role, then some of the behaviors that result will be evolutionarily altruistic. In each case, natural selection must provide the organism with a proximate mechanism that causes the organism to produce the selected behavior. If our behaviors are evolutionarily selfish, should we expect them to be implemented by a mind that is psychologically selfish? And if our behaviors are evolutionarily altruistic, should we expect them to be implemented by a mind that is psychologically altruistic? In short, which proximate psychological mechanisms should we expect to find associated with strictly individual selection on the one hand and with group selection on the other?

In principle, evolutionary altruism can be implemented by psychological altruism *and* by psychological egoism. And evolutionary selfishness likewise can be implemented by the two psychological motivational structures as well. Suppose that natural selection has favored a behavior because the behavior benefits the group (and in spite of the fact that it harms the actor who performs the behavior). One psychological setup that will cause the organism to perform the behavior is for the organism to care only about the welfare of the group. But a quite different arrangement also can do the trick. Let the organism care only about feeling pleasure and avoiding pain, and let the organism be so constituted that how it feels is correlated with how well the group is doing.

The same point applies if we consider a behavior that is evolutionarily selfish. Consider, for example, the behavior of parental care. Parents enhance their own fitness by helping their offspring. Fitness, don't forget, reflects both the organism's chance of surviving *and* its prospects for reproductive success. In our species, as well as in many others, organisms that care for their offspring are fitter than organisms that do not. If individual selection has favored this behavior, and if the proximate mechanism for producing the behavior is to be a mind equipped with beliefs and desires, the question remains of what sorts of preferences that mind will contain. One possibility is that parents are made to care enormously and irreducibly about the welfare of their children. This paramount other-directed preference would mean that parents are altruistically motivated. On the other hand, parents might care only about feeling pleasure and avoiding pain, and be so constituted that their good and bad feelings are correlated with how well their offspring fare. Under this scenario, parents dole out parental care for purely egoistic reasons. I conclude that there is no *a priori* reason why evolutionary altruism must be paired with psychological altruism, and evolutionary selfishness with psychological egoism (Sober, 1993a; Wilson, 1991).

Having separated the evolutionary and the psychological concepts, I now want to consider how they might be connected. Are there evolutionary considerations that help decide whether we are psychological altruists or egoists? The example behavior I will consider is parental care. Human beings provide enormous amounts of parental care, compared with many other species. I'm going to assume that this difference has an adaptive explanation. In saying this, I am not denying that there is variation within our species for the behavior. Child abuse and child neglect are all too common phenomena. I certainly do not deny their existence, although I will say nothing about why they occur. My interest is in between-species differences, not in within-species variation.

If we assume that natural selection played a role in causing human beings, on average, to provide considerable amounts of parental care, the question is whether we should expect natural selection to have done this by providing us with an egoistic or an altruistic motivational structure. That is, we are trying to judge the plausibility of the following two hypotheses about proximate mechanism:

(A) Parents care about their children, not as means to the parents' own happiness, but as ends in themselves.

(E) Parents care only about their own well-being, but they are so constituted that their own well-being is correlated with the well-being of their children.

Our problem is to discover whether natural selection would have favored one of these proximate mechanisms over the other.

In addressing the psychological problem in this way, I am ignoring a rather different avenue of inquiry. It might be suggested that we are sometimes psychological altruists, but that this does not have an adaptive explanation. That is, one might argue that this aspect of our psychology is *evolutionary spin-off* (on which see Sober 1988b). It is like our ability to do calculus or play with hoola hoops; though useless or even deleterious in itself, it was correlated with traits that were advantageous. This is an interesting suggestion, but one that I won't attempt to evaluate here.[9]

5. DIRECT AND INDIRECT STRATEGIES

Once we consider altruism and egoism as alternative mechanisms for implementing a selected behavior, we can situate this problem within a broader evolutionary context. Consider the fact that fruitflies need to live in habitats that are humid. How might natural selection have caused *Drosophila* to move away from arid areas and toward areas of higher humidity? We can identify a *direct* and an *indirect* solution to this design problem (Williams 1966):

(D) Fruitflies have humidity detectors.
(I) Fruitflies have light detectors (not humidity detectors), but darkness and humidity are correlated in the flies' usual habitat.

Of course, (D) and (I) are incompletely specified; we must imagine that each detector is wired to a behavior effector in the right way.

I hope it is clear why I say that (D) is direct and (I) is indirect. If the organism needs to be in a humid locale, the direct solution is to give the organism a detector of that very property. It is more indirect to get the organism what it needs by giving it a detector of some other property that happens to be correlated with the property of interest. (I) is more in the nature of a Rube Goldberg device.

As it happens, (D) is false and (I) is true. Fruitflies have eyes, not humidity detectors. They fly toward darkness; that is how they manage to avoid drying out. So it is false that natural selection always provides direct solutions to design problems. But suppose you didn't already know which of (I) or (D) is correct (and were not allowed to just look and see). What considerations would be relevant to predicting which sort of mechanism the flies will evolve, given that they need a device that steers them away from dryness and toward humidity?

One consideration that is relevant to this problem is the relative *effectiveness* of the two strategies. If darkness and humidity are perfectly correlated, then (D) and (I) will be equally effective. But if the correlation is less than perfect, (D) will be more effective.

Since evolution does not always produce direct solutions, it seems clear that effectiveness is not the only factor influencing which proximate mechanism will evolve. In addition, there is the question of *availability.* We must consider what variants actually were present in the ancestral population in which the trait evolved. Perhaps (D) would have been more effective than (I), but (D) was not available. Natural selection favors the best trait *available;* there is no guarantee that the available traits include all traits that are *conceivable.*

A third consideration, beyond effectiveness and availability, also is relevant. Even if the ancestral population contains both light detectors and humidity detectors, and even if humidity detectors are more effective, it does not follow that natural selection will favor humidity detectors. The complication is that the two traits may have further consequences for the fitnesses of the organisms in which they occur. For example, suppose that humidity detectors impose huge energetic costs on the organism; suppose they require many more calories to build and maintain than do light detectors. This third consideration, which may be lumped into a category called *side constraints,* cannot be ignored. The effectiveness of a device in regulating some particular behavior of interest does not settle how fit the organisms are that have it (Sober 1981).

6. PSYCHOLOGICAL ALTRUISM IS A DIRECT STRATEGY

I hope the bearing of the problem about fruitflies on the problem about human motivation is starting to become clear. Psychological altruism, as described in (A), is a direct solution to the design problem of getting organisms to take care of their offspring. Psychological egoism, as described in (E), is an indirect solution. For fruitflies, we can simply look and see how they manage to find humid spots. However, when it comes to human motivation, the proximate mechanisms underlying parental care are harder to discern. Let us consider (A) and (E) as hypotheses about proximate mechanisms, whose plausibility we can judge in the same way we evaluated (D) and (I).

If a parent's well-being were perfectly correlated with the well-being of his or her children, then (A) and (E) would be equally effective. However, it is perhaps too much of an idealization to imagine that the correlation will be perfect. People occasionally find themselves in bad moods despite the fact that their children are doing well; and sometimes they find themselves in good moods despite the fact that their children are doing badly. If either of these two types of "error" occurs, (A) will be more effective than (E).

What about the availability of the two strategies? Here we may have a *dis*analogy with the fruitfly example. I suspect that *if (E) was available evolutionarily, so was (A).* To see why this conjecture is plausible, let us consider in more detail how (E) would be implemented. According to (E), parents care only about their own well-being, but their well-being is correlated with the well-being of their children. How is this correlation established? Since we are talking about individuals who produce behaviors because of the beliefs and desires they have, we may suppose that these egoists reliably form beliefs about the welfare of their offspring. These beliefs cause the parents to feel good or bad. Since egoists care only about how they feel, they produce actions intended to help their children.

Notice that individuals implementing (E) must be able to form beliefs with the contents "my children are doing well" and "my children are doing badly."[10] This suggests that if (E) is an available mechanism for producing parental care, so is (A). My reason for saying this is based on the following conjecture about how beliefs and preferences are connected: *If people can form beliefs about whether some proposition p is true, they also can form preferences about whether p should be true.* I am conjecturing that belief formation and desire formation are not so rigidly modularized that propositions available to the one are inaccessible to the other.

My suggestion is that the mental equipment needed for someone to be a psychological altruist does not differ in kind from the equipment needed for a person to be an egoist. Granted, the egoist desires instrumentally what the altruist desires as an end in itself; however, the basic belief/desire structure is the same.

These considerations, of course, do not absolutely guarantee that if (E) describes some of the people in an ancestral population, then (A) describes other people in that same population. An accident of history may have excluded (A) but have allowed (E) to make an appearance. What I am suggesting should be understood as a burden of proof argument. I do not see why (A) should have been unavailable evolutionarily. (E) and (A) look similar enough in terms of the basic equipment they require that some special consideration is needed to think that (E), but not (A), was present.

The final question, about side constraints, brings us to the frontier, beyond which there is only *terra incognita.* We know too little about the human mind, and about its evolutionary past, to say much about the side effects that (A) and (E) might have had on the organisms who provided parental care because of them. Within evolutionary theory, this confession of ignorance has different meanings for different biologists. Some biologists may incline to think that it is an acceptable simplification to assume that the trait of interest evolved independently of other traits. Others may regard this as radically implausible. This is an instance of the debate about *adaptationism.* I'm afraid that space does not permit me to discuss the general issue or assess its seriousness for the problem at hand.[11]

In summary, (A) appears to be at least as effective as (E) in delivering parental care. And with respect to the issues of availability and side constraints, there appears to be no reason to think that (A) is less probable than (E). A bold conclusion to draw would be that (A) is more plausible than (E). A more circumspect conclusion would be that there is no evidence favoring (E) over (A).

7. A PUZZLE

In evolution, old characteristics often take on new functions. Sea turtles use their front fins to dig nests in the sand, but the fins did not evolve because they enabled turtles to do this. The fins were in place long before sea turtles emerged from the ocean to lay their eggs. The evolution of psychological motives for parental care is the mirror image of what happened to sea turtles. In the psychological case, a new structure took on an old function. Parental care is a much older characteristic than the cognitive and affective apparatus that provides its proximate implementation in our species. Beliefs and desires were grafted onto a set of behaviors that were at least partially in place before the advent of mind. The point is not that the human mind engendered no behavioral novelties. That is obviously absurd. Rather, my point is that parents were taking care of their offspring long before minds made their appearance.

To understand how mental features were grafted onto a preexisting behavior, it may be useful to compare the evolution of belief with the evolution of desire. No one suggests that our beliefs are solely about our own psychological states. Throw a rock at some people and they will believe *that a rock is approaching;* their belief will not be limited to a description of how they feel. This familiar fact presents an interesting puzzle. Why do beliefs reach out into the world? From the point of view of survival and reproduction, perhaps a belief mechanism would work just fine if beliefs were solely about internal states of the organism that happened to be correlated with external conditions. Solipsism may be *false,* but it is hard to see why solipsism should be *selectively disadvantageous.*

Regardless of how this puzzle is to be solved, it seems clear that we *do* have beliefs about events outside our own minds. The question I now want to pose is this: Why should desire be different? Why should our ultimate preferences be limited to our own psychological states? If beliefs reach out into the world, why should desires not do the same? Perhaps if we could understand why beliefs are about the external world, this might help resolve the puzzle about desire. This is a line of inquiry that I pursue in Essay 11.

8. CONCLUSION

I have argued that there is no evolutionary reason for preferring the hypothesis of psychological egoism. Neither effectiveness, nor availability, nor side constraints offers us a reason to think that human beings take care of their children for purely egoistic reasons. In the light of this conclusion, it is a curious fact that psychological egoism continues to dominate thinking in psychology and the social sciences. The very idea that people have irreducibly other-directed preferences, and that these sometimes are stronger than whatever self-directed preferences they have, is often viewed as naive. The implicit attitude of many theorists seems to be that if a range of behaviors *can* be explained by the hypothesis of egoism, then it *should* be so explained. The fact that the hypothesis of altruism *also* could explain the behavior is dismissed or ignored. However, it is worth asking why egoism should be accepted as the explanation of a behavior if the behavior also is consistent with the hypothesis of altruism. In the background of this bias in favor of egoism is the idea that egoism is the more parsimonious hypothesis (Batson 1991). Since we already have reason to think that *some* motives are egoistic, it is more parsimonious to suppose that *all* are ultimately egoistic.[12] I believe that there is less to this parsimony argument than might at first appear. If both hypotheses can explain the behavior, I don't see why parsimony provides much of a reason for favoring egoism.[13] In my opinion, psychological egoism does not deserve to be the default hypothesis that we blithely assume until we are forced to think otherwise.

Reinforcing the inclination of many social scientists to accept a selfish view of human nature is the belief that evolutionary theory leads to this very conclusion. But, in fact, evolutionary biology entails no such psychological consequence. Even if various aspects of our behavior (e.g., parental care) are *evolutionarily* selfish (i.e., were molded by individual selection), it is not immediately obvious why *psychological* egoism should have been the proximate mechanism that evolution provided to implement the behavior. Quite the contrary; it is rather puzzling why creatures whose fitness depends on the welfare of others and who are capable of forming beliefs and preferences about a wide range of propositions should be so constituted that they care only about themselves. Psychological egoism is not at all what one would expect from the point of view of evolutionary biology.

NOTES

1. There is a fine point of disanalogy here: If a behavior is an example of psychological altruism, then it *must* proceed from motives of a certain sort. However, if a behavior is an example of evolutionary altruism, it may simply have arisen yesterday by mutation; it *need not* have been the result of group selection. Nonetheless, if further assumptions are adopted, the hypothesis of group selection is *de facto* a commitment that follows from the label of evolutionary altruism.
2. For example, I won't discuss whether kin selection and game theoretic interactions are properly regarded as varieties of individual selection or of group selection. On this, see Sober (1992a, 1993b).
3. Many philosophers seem to believe that Joseph Butler refuted egoism, at least in its hedonistic form. I do not. My objections to Butler's argument are given in Sober (1992b).
4. As used here, "satisfaction" does not name a sensation, but denotes a logical relation between an outcome and a preference. If I want it to rain tomorrow and rain occurs, my desire has been satisfied, even if I never learn that it has rained; in this case, the desire is satisfied even though I receive no pleasure. My thanks to Daniel Hausman for impressing on me the importance of emphasizing this point.
5. Here and in what follows my description of what agents will do in a given situation should be understood as shorthand for what they will do, given their *beliefs* about the situation.
6. Notice that the taxonomy I have proposed entails that an egoist can care about the welfare of others. Moderate egoists prefer that others be better off rather than worse, quite apart from the effect this might have on their own well-being. For them, an interest in the welfare of others is irreducible. I nonetheless classify this motivational structure as an instance of egoism because such individuals are never prepared to sacrifice their own welfare for the sake of another's. They are egoists because their interest in the welfare of others is too weak to counteract the concern they have for themselves. They are egoists because they say "me first."
7. Egoism, understood in terms of thesis (E1), makes the following claim about the preferences an agent will have in a given choice situation:

> For each other-directed preference O, there exists a self-directed preference S, such that the agent has preference O only because O is an instrument for satisfying preference S.

We now can say that an agent has O only because it is an instrument for obtaining S precisely when O and S are related as follows:

	+O	−O
+S	4	4
−S	1	1

For example, extreme egoists may prefer that their children do well rather than badly, but their reason for doing this is simply that the welfare of their children is correlated with whether they themselves feel good or bad.

8. Of course, perfect precision isn't necessary. A device can make both type-one and type-two errors and still evolve, if it is superior to the other devices with which it competes. This is discussed in more detail in Essay 3.
9. Singer (1981) develops an argument of this type. His idea is that "if there are advantages in being a partner in a reciprocal exchange, and if one is more likely to be selected as a partner if one has genuine concern for others, there is an evolutionary advantage in having genuine concern for others" (p. 44). Notice that Singer is proposing a correlation between psychological altruism and being chosen as a partner in exchanges. It also is worth noting that Singer's argument relies on there being a behavioral difference between egoism and altruism. My analysis will be compatible with their being behaviorally indistinguishable.
10. The concept "my children" isn't essential. The point is that the beliefs must have contents that are *about other people.*
11. For discussion pro and con see Maynard Smith (1978), Gould and Lewontin (1978), Orzack and Sober (1994), and Sober (1993b).
12. Batson (1991, chapter 4) suggests that we should prefer the egoistic hypothesis on grounds of parsimony if there were no behavioral evidence favoring altruism over egoism. However, Batson argues that there *is* behavioral evidence that supports the hypothesis of altruism.
13. I develop more general reasons for being skeptical of such parsimony arguments in Sober (1988a) and in Essay 7.

REFERENCES

Batson, D. (1991). *The Altruism Question: Towards a Social-Psychological Answer.* L. Erlbaum.

Darwin, C. (1871). *The Descent of Man, and Selection in Relation to Sex.* Reprinted, Princeton University Press, 1981.

Dawkins, R. (1976). *The Selfish Gene.* Oxford University Press.

Gould, S., and Lewontin, R. (1978). The Spandrels of San Marco and the

Panglossian Paradigm: A Critique of the Adaptationist Programme. *Proc. Royal. Soc. London 205:* 581–98.

Maynard Smith, J. (1978). Optimization Theory in Evolution. *Annual Review of Ecology and Systematics 9:* 31–56.

Mayr, E. (1961). Cause and Effect in Biology. *Nature 134:* 1501–6.

Orzack, S., and Sober, E. (1994). Optimality Models and the Test of Adaptationism. *American Naturalist 143*:361–80.

Singer, P. (1981). *The Expanding Circle: Ethics and Sociobiology.* Farrar, Straus, and Giroux.

Sober, E. (1981). The Evolution of Rationality. *Synthese 46:* 95–120.

Sober, E. (1984). *The Nature of Selection: Evolutionary Theory in Philosophical Focus.* MIT Press.

Sober, E. (1988a). *Reconstructing the Past: Parsimony, Evolution, and Inference.* MIT Press.

Sober, E. (1988b). What Is Evolutionary Altruism? In B. Linsky and M. Matthen (eds.), *New Essays on Philosophy and Biology* (*Canadian Journal of Philosophy* Suppl. Vol. 14), University of Calgary Press, pp. 75–99.

Sober, E. (1989). What Is Psychological Egoism? *Behaviorism* 17: 89–102.

Sober, E. (1992a). The Evolution of Altruism – Correlation, Cost, and Benefit. *Biology and Philosophy 7:* 177–88.

Sober, E. (1992b). Hedonism and Butler's Stone. *Ethics 103:* 97–103.

Sober, E. (1993a). Evolutionary Altruism, Psychological Egoism, and Morality: Disentangling the Phenotypes. In M. Nitecki (ed.), *Evolutionary Ethics,* SUNY Press, pp. 199–216.

Sober, E. (1993b). *Philosophy of Biology.* Westview Press.

Williams, G. (1966). *Adaptation and Natural Selection.* Princeton University Press.

Wilson, D. (1991). On the Relationship Between Evolutionary and Psychological Definitions of Altruism and Egoism. *Biology and Philosophy 7:* 61–68.

2

Why not solipsism?

1. THE QUESTION

Solipsism poses a familiar epistemological problem. Each of us has beliefs about a world that allegedly exists outside our own minds. The problem is to justify these nonsolipsistic convictions. One standard approach is to argue that the existence of things *outside* our own sensations may reasonably be inferred from regularities that obtain *within* our sensations. Certain experiences, which I will call tiger sounds and tiger visual images, exhibit a striking correlation. We can explain the existence of this correlation by postulating an entity that is a cause of both. If there were tigers, it would be no surprise that certain sights and certain sounds tend to co-occur. Our rejection of solipsism can thus be justified by appeal to an abductive argument; we advance an inference to the best explanation conforming to the pattern that Reichenbach (1956) called *the principle of the common cause* (Salmon 1984; Sober 1988).[1]

The epistemological problem posed by solipsism is an old chestnut. A quite different problem – one less often noticed, let alone addressed – is the problem of explanation. *Why* do we have beliefs about the world outside our own minds? To reply that it is in our nature to "reify" may be true enough, but it merely postpones the question at hand. *Why* are our minds so constituted that we spontaneously think in nonsolipsistic terms?

This question is evolutionary in character. In the lineage leading to

I am grateful to Don Moskowitz and Greg Mougin for their detailed criticisms of various ancestors of the present paper. I also thank Fred Dretske, Berent Enç, Malcolm Forster, Martha Gibson, Paula Gottlieb, Eric Saidel, Mark Sainsbury, Alan Sidelle, Dennis Stampe, and Leora Weitzman for useful discussion. In addition, I benefited greatly from discussing this paper at University of Minnesota, at University of Michigan, and at University of Miami.

human beings, intentionality made its appearance. The ability to form beliefs, desires, and other representations with semantic content gradually came into being. My question is not the origin of content, but the origin of nonsolipsistic content. Given the evolution of beliefs that purport to describe *something,* why did those beliefs come to describe things outside the subject's own mind?

I want to address a part of this evolutionary question by considering the issue of functional utility. What good does it do a creature with a mind to have representations that are nonsolipsistic? What advantage does this arrangement have over solipsism? Couldn't a creature get around in the world just as well by forming representations solely about its own mental states? If tiger beliefs are useful because they make us run away when tigers are present, couldn't the same behavior be wired to the solipsistic belief that tiger images or tiger sounds are present?

The issue of adaptive value is only part of the larger problem of providing an evolutionary explanation for the trait of interest. Describing the trait's functional utility leaves open how important that functional utility was in the trait's evolution. Perhaps the trait is present for reasons wholly distinct from its functional utility; or perhaps functional utility was only a small part of the constellation of factors that caused the trait to evolve. Although identifying the functional import of a trait isn't the whole explanation of why the trait evolved, it is an important issue to address. If we wish to ask whether a trait evolved because it was beneficial, we first must figure out whether the trait conferred a benefit and what that benefit was. In this respect, the problem of nonsolipsistic content is no different from the problem of explaining any other evolutionary novelty (Orzack and Sober 1994; Sober 1993).

To organize my inquiry, I need to distinguish three questions about solipsism. Two have already been mentioned:

(S) In virtue of what does a given belief purport to describe the world outside the subject's mind?
(E) Assuming that some beliefs do describe the world outside the subject's mind, how does the subject ever *know* (or have a *rational justification for thinking*) that those beliefs are true?
(F) Why is it *useful* to an organism for it to have beliefs that describe a world outside its own mind?

(S) is a semantic question. It asks us to describe the nature of nonsolipsistic intentionality. This task differs from that of providing a causal explanation for why organisms have beliefs about the world outside of themselves. In just the same way, asking what the nature of water is differs from asking why lakes contain water. The question about lakes is a request for causes; the question about the nature of water is not. You can answer the question about the nature of water by saying that water is made of H_2O molecules; however, being made of H_2O molecules doesn't *cause* something to be made of water (at least not if cause must precede effect).

(S) is the most fundamental question of the three. The others presuppose that aboutness makes sense and that the distinction between solipsistic and nonsolipsistic representations is an intelligible one. (E) is the traditional problem of epistemic justification. (F) is the question of function. I will try to answer this third question, but not the other two.

My question of why we are not solipsists may seem odd, given my conviction that there exists a world outside the mind and that people are quite reasonable in thinking that this is so. When people have *false* beliefs, the question of functional utility has a more obvious bite. This is why atheists over the years have found such interest in explaining the persistence of theism. But if tigers exist as mind-independent entities and it is reasonable to think that this is so, what more is there to say about the usefulness of including this proposition in one's stock of beliefs?

In point of fact, there is precedent for asking evolutionary theory to explain why we have beliefs that we happen to regard as both true and reasonable. Darwin and Wallace debated the question of how one should explain the human inclination to form beliefs about purely theoretical matters (Gould 1980). What survival value could there have been for our ancient ancestors to possess the cognitive abilities that later allowed mathematical science to flourish? Even if one assumes that relativity theory is true and well-supported by evidence, the question remains of explaining why human beings evolved the capacity to think up relativity theory. By the same token, even if it is both true and reasonable to believe that a world external to the mind exists, the question remains of explaining why human beings evolved the inclination to think in such nonsolipsistic terms.

One way to approach the functional question is to link it to the semantic problem. The strategy here is first to provide a general theory that says when a belief has the content *object o has property F,* and then to show that it follows from that theory, or from that theory when supplemented by unproblematic further assumptions, that many of the beliefs we have are nonsolipsistic – i.e., that the *o*'s that figure in their content purport to denote objects outside the mind.

There are three reasons why I find this "direct" approach to the functional question unattractive. First, there is no theory of content that I regard as adequate. Second, I think it is implausible that nonsolipsistic conclusions should follow from the nature of belief or from the nature of semantic content. After all, *some* of our beliefs are about what goes on inside our own minds. If some are like this, why couldn't they *all* be? Some will find this question naive; however, I am not persuaded that the question is confused. I don't offer this point as a positive argument that all of our beliefs *could* be solipsistic; I merely remark that negative arguments to the contrary have never struck me as persuasive.[2]

The third reason for eschewing a direct approach to the functional question is, I think, the most important. Suppose, for the sake of argument, that the nature of belief and the nature of semantic content somehow entail that beliefs must have nonsolipsistic contents. Even if this were true, the functional question would remain to be answered. For we still would need to ask why beliefs and semantic contents were more useful to the organism than were beliefs* and contents*, where these starred items are just like their unstarred counterparts, save for the fact that they are solipsistic. Essentialism is no way to solve or dissolve problems of functional utility. If one is asked what good it does for blood to transport oxygen, it is irrelevant to reply that blood is, by definition, an oxygen transporter.

So, for these three reasons, I will attempt to answer the functional question by pursuing an *in*direct strategy. That is, I'll try to describe the functional utility of nonsolipsistic belief without resting my case on a full-blown theory of what belief or semantic content is. Of course, I won't be able to get by without making *some* assumptions about these mentalistic categories. But my goal is that these assumptions should be fairly innocuous; they should be no more than sensible requirements that any adequate theory of belief and content can be expected to satisfy.[3]

If our task is to assess the functional utility of solipsism and nonsolipsism, the first problem is to clarify what these two phenotypes amount to. Solipsistic organisms have beliefs that are solely about their own experiences; these beliefs make no reference to a world that exists outside the mind. A solipsist may think something like *I am experiencing a bitter taste* or *bitterness is now occurring.*[4] However, the solipsist cannot think *this tastes just like the lettuce I ate yesterday.*

I do not limit solipsism to solipsism of the present moment. Solipsistic organisms form beliefs about present *and* past experience, and they make predictions about future experience as well. However, solipsists need not realize that they are solipsists; like their nonsolipsistic counterparts, they need not form a belief that is about the character of their own beliefs. Such second-order representations are rather sophisticated achievements and are not part of what I wish to discuss here.

Solipsists form beliefs that are composed from a limited stock of concepts. Each of the nonlogical terms must (purport to) refer to a kind of experience; this is why reference to lettuce, as ordinarily understood, is prohibited. It follows from this rather traditional understanding of solipsism that certain kinds of inference are prohibited. Solipsists can make inductions about regularities that obtain in the course of their experience, but they cannot postulate the existence of something that exists outside of their experience. In short, solipsists make inductions; they do not reason abductively.

2. AN ANSWER

The answer to the functional question that I want to defend is based on an idea developed by Fred Dretske (1986). The answer (or part of it) is his, but Dretske used this idea to answer a question that differs from the one at issue here. Dretske presented his idea as a solution to what I am calling the semantic problem (S). I don't think his proposal succeeds in this context, for reasons that I will explain. However, it does handsomely as an answer to the functional question (F), or so I will argue.

Consider two stages in an organism's lifetime, which I'll begin by describing in as neutral a way as I can. During the first stage, tiger images tend to produce tokenings of belief *A*, which then combine with other beliefs and desires to generate behaviors. Suppose that the organism never encounters an antelope during this first stage. During the second stage, the organism continues to form beliefs of type *A* when tiger images occur, but it begins to token beliefs of type *B* when (bloodied) dead antelopes are present. Beliefs of this type then join with other beliefs and desires to produce behaviors. So, of the two patterns described below, the first is present at stage one, whereas both are present at stage two:

tiger ⟶ tiger image ⟶ belief *A* ⟶ behavior
dead antelope ⟶ dead antelope image ⟶ belief *B* ⟶ behavior

Part of my desire to begin with as neutral a description as I can is that I want to leave it open in my formulation of the problem whether *A* and *B* are, in fact, the same belief.

I have not said enough about these two beliefs for one to be sure what their contents are. Nevertheless, I want to consider two hypotheses:

(NON) Belief *A* and belief *B* have the same content; both mean *a tiger is present.*

(SOL) Belief *A* means *a tiger image is present;* belief *B* means *a dead antelope image is present.*

(SOL) attributes solipsistic contents to the organism's beliefs; (NON) attributes nonsolipsistic content. Our task is not to answer the semantic question of explaining why one of these hypotheses is *true;* we are addressing the functional question of why it should be more *useful* to the organism for it to be built so that (NON), rather than (SOL), is true.

My answer to this functional question has two components. First, notice that (SOL) involves a difference in representational content, whereas (NON) says that the two beliefs have the same content. I want to argue that this is no accidental feature of my choice of examples, but reflects something fundamental about the difference between solipsism and nonsolipsism. Second, I'll argue that this difference between (NON) and (SOL) makes a functional difference. There is an advantage to having phenomenologically different experiences trigger beliefs that have the same content.

(SOL) postulates a difference in content; (NON) does not. Let's consider several objections that seek to show that this asymmetry is entirely incidental to the difference between solipsism and its opposite. These objections maintain either that solipsistic contents can plausibly be postulated that involve a univocal treatment of the two beliefs, or that the nonsolipsistic approach must postulate different contents for *A* and *B.*

First, consider the solipsistic suggestion that one should attribute to both *A* and *B* the disjunctive content *a tiger image or a dead antelope image is present.* My reply is that this suggestion isn't plausible as a thesis about the content of *A* during the first stage of the organism's life, since the individual then had no contact whatever with antelopes. Of course, it would be nice to have a theory of content that underwrites this judgment. I can't offer one. However, as mentioned above, this isn't an objection to the thesis I am advancing. I contend that my reply to this objection involves a plausible supposition, which any adequate theory of content should be expected to accommodate.[5] When my children were young, they had the concept *dog,* but they did not have the concept *dog or electron microscope;* the organism in my example, at stage one, is in precisely the same boat.

A related objection is that *A* and *B* both have the solipsistic content

I am being appeared to tigerly. My response is to question what it means to be "appeared to tigerly." It can't mean "being appeared to in the way I normally am when a tiger is present," since this content isn't solipsistic. The phrase must label a kind of experience, and do this without reference to the external world. So the term must subsume tiger visual images, dead antelope images, tiger sounds, the sound of someone saying "oh gosh, there's a tiger in that bush!", and whatever other experiences I might have that would make me think that a tiger is nigh. Surely these experiences have nothing in common *phenomenologically.* What they do have in common is that each makes me think that a tiger is present.

My objector might contend that "being appeared to tigerly" just codes for the disjunction "tiger visual image, or dead antelope image, or. . . ." I won't object to the fact that the ". . ." is a promissory note. My answer is that I don't think that a person who has never experienced an antelope has this concept. This restates a point already made in connection with the solipsistic disjunction "tiger image or dead antelope image." If "being appeared to tigerly" is short for a specific solipsistic disjunction, then it can't capture the content of both *A* and *B.*

The next objection I want to consider is that the solipsistic hypothesis must postulate the same content for both beliefs, because both beliefs produce a certain bodily movement ("running away," say), and this resulting movement is what gives the state its content; at both stages, the organism believes *I should run.* To block this suggestion, I will stipulate that there is no single behavior that always occurs when *A* is tokened, nor any that always occurs when *B* is tokened. Beliefs of type *A* sometimes lead to running away, sometimes to remaining motionless, sometimes to pushing tiresome people in a certain direction, etc. In short, I am assuming that a crude behaviorist rendering of the contents of belief states isn't acceptable. Behaviors triggered by beliefs are not like fixed action patterns. The relation of nearby tigers to running away in humans isn't like the connection of empty burrows to inspection behavior in the wasp *Sphex* (Dennett 1978).

My claim is that *A* and *B* will have distinct contents if they are construed solipsistically, but that they can have the same content, if they are construed nonsolipsistically. So far, I have considered objections to the link I see between solipsism and difference of content; now I want to consider an objection to the thesis that it is reasonable to regard *A* and *B* as having the same content, if they are construed nonsolipsistically.

The objection maintains that *A* and *B* must have different contents because *A* can be triggered by tiger images, whereas *B* can be triggered by dead antelope images. In rejecting this claim, I am assuming that verificationism, and theses like it (such as operationalism, on which more below), are false. The fact that new methods of detection are added to an organism's repertoire doesn't *automatically* mean that new and old beliefs must differ in content. It is possible (though not inevitable) for new methods to be brought into play for triggering the same old belief contents. I take it that no theory of meaning should say this cannot happen.[6]

This completes my defense of the claim that a solipsistic rendering of the contents of *A* and *B* must involve a difference in belief content, whereas a nonsolipsistic interpretation need not. I now turn to the functional significance of this difference. Why is it better for the organism to assign the same belief content to *A* and *B?* Obviously, this sort of univocity isn't a categorical imperative. Indeed, it is altogether plausible to imagine that the organism we are talking about, even if it is nonsolipsistic, will trigger slightly different *sets* of beliefs when a tiger image occurs than it will when a dead antelope image is present. The crucial question is why there should be an advantage in having a *common element* in the sets of beliefs triggered on those two types of occasion.

To address this issue, we need to distinguish the general beliefs an agent has, which often persist from one moment to another, from the singular beliefs that get tokened when sensory stimuli are one way, but are often cancelled when they are another. *Tigers are dangerous* is an example of the first sort of belief; *there is a tiger nearby* is an example of the second. In the example I have been discussing, *A* and *B* are singular beliefs. Such beliefs give rise to action only when they are conjoined with what I am calling general beliefs. If a sensory stimulus makes me think *there is a tiger nearby,* this will lead me to run away only if I have in hand such general beliefs as *being near tigers is dangerous* and *running away from a tiger reduces danger.*

If an agent repeatedly tokens beliefs that have the content *a tiger is present,* the agent can exploit a stock of general beliefs that give advice about what to do in that kind of circumstance. Keeping the content of *B* the same as the content of *A* permits the agent to continue to exploit this fund of information. In contrast, if *B* has a content that differs from the one assigned to *A,* the wealth of information that helped decide what to do when belief *A* was tokened no longer is relevant as guidance for what to do when the agent subsequently believes *B.*

Consider the mental life of the organism we are discussing if its beliefs are solipsistic. During the first stage in the organism's life, the organism knows what to do when it believes that *a tiger image is present.* But the general beliefs that give guidance here are likely to be of no avail during stage two, when the organism starts to form the belief that *a dead antelope image is present.* If the content of *B* differs from the content of *A,* the organism is apt to be left high and dry at the beginning of this second stage. The organism no longer knows what to do, but needs to start from scratch, developing a new fund of general beliefs that are useful when this new belief content gets tokened.

It isn't that the solipsist can never figure out that tiger images and dead antelope images have something interesting in common. Perhaps repeated experiences of each will cause a solipsistic organism to realize that behaviors appropriate to the former also are appropriate to the latter. But this discovery requires a long run of samples, one that predatory tigers may summarily nip in the bud. The great thing about the nonsolipsist is that its belief in tigers provides it with a category that even the *first* exposure to a dead antelope may trigger. The first dead antelope is probably an interestingly novel type of experience for the solipsist, but is more likely to be a sign of danger to the nonsolipsist. This difference can make all the difference.

3. THE DRETSKE CONNECTION

As already mentioned, Fred Dretske (1986) has addressed the semantic problem of explaining what it is about the state that an organism occupies that endows the state with a particular semantic content. Dretske (1986, 1988) thinks that semantic content is constrained by two considerations. First, if a state means that *a tiger is present,* then a tiger must be present whenever the state is tokened during an initial period of time during which the state acquires its content. Let us describe this first constraint by saying that *content entails correlation.* Dretske's second constraint is that a state means that *a tiger is present* only if the state has the function of indicating the presence of tigers. According to Dretske, *content entails function.*

This second necessary condition on content, Dretske argues, is not enough, because there are numerous ways to describe a state's function. Tiger detectors have the function of indicating the presence of tigers, but it also is true that they have the function of indicating the presence of dangerous things. Functional considerations aren't enough to uniquely determine content, Dretske argues.

Dretske suggests that a third constraint on content can be found by considering an organism that learns. If an organism initially is caused to go into a particular physical state by tiger images, and then acquires the ability to have dead antelope images send it into that state as well, what is the content of this state? Dretske assumes that the state retains its function through time; he takes this sameness of function to ensure that the meaning of the state remains unchanged. He concludes that the state's single invariant content must be nonsolipsistic; the state must mean that *a tiger is present.*

Let me explain some of the respects in which my use of this diachronic consideration differs from the use that Dretske makes of it. First, I do not assume that Dretske's first two constraints on meaning are correct. I do not require that content entails correlation, nor that content entails some claim about function. Second, I do not think that Dretske has defended his assumption that the function of the physical state must be the same across the two temporal intervals. It is not inevitable that a physical state should retain the same function through an organism's lifetime. A bird's plumage may have the function of attracting mates in one season, but of concealing the bird from predators in another.[7]

An additional difference between Dretske's use of the diachronic consideration and the use I make of it is that we are addressing different questions. The semantic question (*S*) is what Dretske aims to answer, whereas my interest is in the functional question (*F*). And, of course, a crucial element in my proposal is the informational utility of having two states possess the same content; this idea does not figure in Dretske's proposal.

One last contrast is also worth drawing. Dretske (1988, pp. 89–92) extracts anti-innatist consequences from his account of semantic content. Dretske interprets his model as describing not just *one* pathway by which states can come to have semantic content; he believes that semantic content arises *only* from a learning process. My functional argument has no such implication. I have tried to describe *one* advantage that nonsolipsistic content provides; even if that advantage were available only for beliefs that are learned, my proposal would not rule out the possibility that there are *other* advantages entailed by nonsolipsistic contents that accrue to beliefs that are *not* learned. In addition, it is consistent with my proposal that innate beliefs should be nonsolipsistic even if this confers no advantage whatever.

In spite of these different stances that Dretske and I take toward the issue of innateness, there does exist one anti-innatist assumption that underlies my argument. I have construed the solipsistic hypothesis as entailing that when the organism first sees a dead antelope, it probably will have no inkling as to what behaviors it should produce. That is to say, the disadvantage I have attributed to solipsism depends on the assumption that organisms do not always know instantaneously and innately what behaviors to produce when various sensations occur. According to my scenario, selection will favor solipsism over nonsolipsism only when innate structures are an insufficient guide to effective behavior.[8]

4. OPERATIONALISM AND THE THEORETICIAN'S DILEMMA

A detailed parallel can be drawn between the problem of solipsism and the problem in philosophy of science posed by operationalism. Operationalism says that theoretical magnitudes are defined by the methods used to measure them. An object's temperature is, by definition, what a thermometer says; an individual's level of intelligence is, by definition, what an IQ test says.

Operationalism is a theory of meaning that is manifestly false to the practice of science. It contradicts the obvious fact that methods of measurement can be inaccurate. Thermometers and IQ tests are perfectly capable of *mis*measuring the magnitudes with which they are associated. It is a substantive scientific problem to discover when and why measuring devices work well, and when and why they fail to do so.

A second defect in operationalism is that it individuates theoretical magnitudes too finely. According to this theory of meaning, it is impossible that two different measuring procedures should both provide measurements of one and the same theoretical magnitude. Instead of saying that procedure P_1 and procedure P_2 both provide measurements of *temperature,* the operationalist requires us to say that P_1 measures $temperature_1$ and P_2 measures $temperature_2$, where $temperature_1$ and $temperature_2$ are quite distinct theoretical magnitudes.

To see what is wrong with this requirement, suppose we were to apply each of these two measurement procedures to a series of different objects. We find that the results of P_1 and P_2 are extremely well correlated; higher than average values on the one procedure tend to be associated with higher than average values on the other. The natural way to explain this correlation is to postulate a common cause. P_1 and P_2 are correlated because each measures the same thing – namely *temperature.* Operationalism prohibits us from drawing this sensible conclusion, and thus requires us to regard the correlation of P_1 and P_2, and of *temperature*$_1$ and *temperature*$_2$, as inexplicable mysteries. However, there seems to be no good scientific reason why we should decline to explain the correlation in the way just sketched.

Operationalism, like solipsism, is a kind of epistemic puritanism. It demands abstinence, where more liberal philosophies permit us to indulge. But just as in the case of solipsism, we can ask a functional question about operationalism as well as an epistemic one. I have just sketched the standard *justification* for rejecting operationalism as an account of the meaning of theoretical terms in science. A question distinct from this problem in epistemology and semantics is *why* science developed in such a way that operationalism turned out to be false. Why is it *advantageous* for scientists to use theoretical terms so as to violate the tenets of operationalism? What's in it for them?

The informational argument presented earlier in connection with solipsism applies to the functional problem posed by operationalism. If scientists continue to use a univocal concept of *temperature,* they can exploit an accumulating wealth of general information concerning how singular beliefs of the form *this object has a temperature of x* should be interpreted. However, if there is no such univocal concept, but merely a sequence of unrelated concepts *temperature*$_1$, *temperature*$_2$, etc., then the general information pertinent to interpreting singular judgments of the form *this object has a temperature*$_i$ *of x* is less likely to be brought to bear on subsequent judgments of the form *this object has a temperature*$_j$ *of y* (for all $i \neq j$). Operationalism and solipsism entail similar sorts of semantic fragmentation. It makes sense that the default setting of our use of terms should violate the requirements of both philosophies.[9]

The issues I have described concerning solipsism also bear on what Hempel (1965) called the *theoretician's dilemma.* Craig's theorem and the Ramsey sentence technique both show how it is possible to capture the predictive power of a scientific theory without resorting to theoretical commitments. Reference to electrons can be eliminated from the electron theory, while retaining all of the observational predictions that the electron theory entails. The point to notice is that these logical results concern the *static* properties of a theory, but do not address *dynamic* characteristics of theory change. If we ask why theoretical terms should not be eliminated, an answer may be obtained by considering the advantages such terms afford to a theory as it develops through time. The predictive utility of theoretical terms is to be found in their capacity to allow theories to grow; static snapshots of a theory render this utility invisible.

To make this idea more concrete, I want to describe a simple example that shows how the functional argument applies to the problem posed by Craig's theorem. Consider two theories T_1 and T_2. V and W are theoretical terms in both; G_1 and H are observational terms in both theories as well. However, G_2 is an observational term that occurs only in T_2. Here are the two theories, each divided, as Craig's theorem requires, into its purely theoretical part, its purely observational part, and the part that provides bridge principles connecting the theoretical and the observational vocabularies:

Theoretical:	If V then W.	
Bridge:	If G_1 then V. If W then H.	(T_1)
Observational:	If G_1 then H.	

Theoretical:	If V then W.	
Bridge:	If G_1 then V. If W then H.	(T_2)
	If G_2 then V.	
Observational:	If G_1 or G_2, then H.	

The Craigian replacement (C_1) of T_1 is just the observation statement "If G_1 then H." The Craigian replacement (C_2) of T_2 is the observation statement "If G_1 or G_2, then H." In each case, the theoretical vocabulary does not augment the set of observational consequences that the theory possesses, over and above what is entailed by the Craigian replacement. So what contribution does theoretical discourse make with respect to the goal of predicting observations?

An answer may be obtained by considering how T_1 might *grow*. If we believe the full theory T_1, we may be more likely to consider the possibility that G_2 and G_1 have the same predictive significance. But without that theoretical perspective, it may be less easy to get T_1 to grow into T_2.[10,11] What we have here is a restatement of the familiar point that theories allow us to see that observationally dissimilar situations sometimes have the same underlying theoretical significance.

Quite apart from the answer I have given to the functional question posed by Craig's theorem, doubts can be raised about the very question I have formulated. If it is entirely rational to believe that electrons exist (as I grant), isn't that reason enough to include discussion of electrons in our scientific theories? If this were correct, then no functional question would remain, once the epistemological problem is solved.

Just as in the case of solipsism, I churlishly resist this simple solution. There are lots of true and reasonable propositions that do not get included in a given scientific theory. Physicists, I have noticed, do not list my telephone number when they provide theories concerning atomic structure. One explanation of why they slight me in this way is that my telephone number would be irrelevant to predicting the phenomena that such theories aim to handle. However, Craig's theorem raises precisely the same question about theoretical terms. Why should we include such elements in a theory, when they in fact are irrelevant to the task of predicting observational phenomena?

Philosophers have often been tempted to meet this challenge by replying that prediction isn't the only goal of science. Even if talk of electrons doesn't augment the predictive power of the electron theory, we talk about them anyway, because we are interested in explanation and understanding for their own sakes. Although there is merit to seeing science as having an irreducible interest in theory, this response ignores the question of whether theoretical statements help us make better predictions.[12] My diachronic argument is intended to describe how theoretical terms make an ineliminable contribution to our ability to interpret "new" observations by allowing us to embed them in a preexisting theoretical context.[13]

An example may help clarify the point I am making, and to show that it is fundamentally *psychological,* not *logical,* in character. A real-life physician thinks about both diseases and symptoms. But imagine a Craigian physician, who thinks only about symptoms. How will these two individuals handle a *novel* symptom? Their problem is to predict what further observations the new symptom indicates. If the symptom is not observationally similar to any previously observed symptom, the Craigian physician will probably draw a blank. The symptom will be duly noted and in the course of time it may be found to be correlated with other observable features. Real-life physicians, on the other hand, have additional resources beyond "wait and see." They have a catalog of diseases to consider, and thinking about these can help them formulate hypotheses about the character of the novel symptom. So my claim is not that the Craigian physician cannot ever understand the novel symptom whereas the real-life physician will immediately reach the right explanation. Rather, my point is that the difference in resources will *probably* make a difference in the *efficiency* with which the problem is solved.

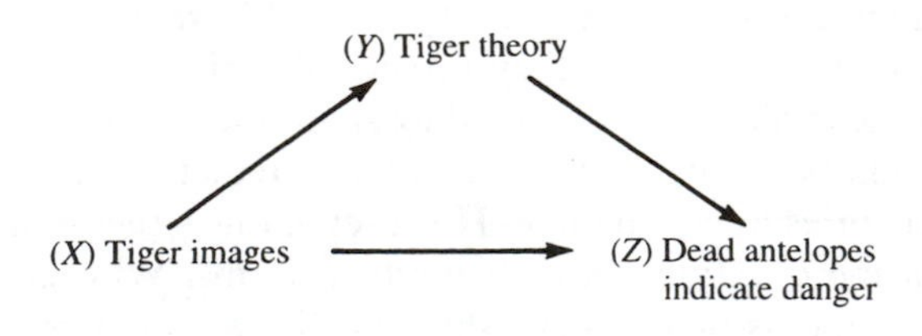

Figure 2.1

5. THE TRANSITIVITY OBJECTION

I hope that the previous section makes a convincing case for my claim that the problem of solipsism and the problem posed by Craigian elimination are structurally the same. For those already disposed to think that theory makes an indispensable contribution to prediction, my argument for the utility of abandoning solipsism should have some force. But another quite consistent reaction to this analogy is to question both the argument about solipsism *and* the argument about the usefulness of theory. In this section, I want to articulate one version of this type of objection.

I have argued that the organism's experience with tiger images during stage one allowed it to postulate the existence of tigers (and a concomitant tiger theory that fleshes out what this postulate amounts to). I then argued that the theory of tigers provided the organism, at the dawn of stage two, with a format that allowed it to see that a dead antelope might mean that danger is at hand. The objection takes the form of a claim of dispensability: if the nonsolipsist can postulate the existence of tigers at stage one, where this postulate provides help in interpreting the appearance of dead antelopes in stage two, then the solipsist will also be able to interpret the appearance of dead antelopes, but without having to pass through the nonsolipsist's construction of a tiger theory.

This objection is depicted in Figure 2.1. The arrow means "supports." The claim is that if X supports Y and Y supports Z, then X supports Z. The conclusion drawn is that talk of tigers is dispensable, as far as the task of interpreting the newly perceived dead antelope is concerned.

Although it is by no means automatic that the support relation is transitive, it can be rendered transitive by adopting some background assumptions. So let us assume, for the sake of argument, that transitivity holds for the kind of case that my tiger story exemplifies. This amounts to the concession that if the nonsolipsist's theory of tigers justifies the claim that dead antelopes indicate danger, then the solipsist's experience with tiger images must, by itself, also justify the same claim:

(L) If X supports Y and Y supports Z, then X supports Z.

The objection we are considering takes the truth of (L) to show that the tiger theory is dispensable. The nonsolipsist makes the first argument listed below and the solipsist makes the second:

$$\begin{array}{l} X \\ X \text{ supports } Y \text{ and } Y \text{ supports } Z \\ \hline\hline Z \end{array} \qquad \text{(I)}$$

$$\begin{array}{l} X \\ X \text{ supports } Z \\ \hline\hline Z \end{array} \qquad \text{(II)}$$

The double lines indicate that the arguments are not supposed to be deductively valid. The objection we are considering says that if the premisses in (I) are true and support the conclusion, the same can be said of (II). If so, all talk of the intervening variable Y may be eliminated.

My reply to this objection is that the epistemological points just granted do not touch the thesis I am advancing. Although talk of tigers may be *epistemically* eliminable (in the sense just defined), it may nonetheless play an indispensable role in getting the organism (O) to *consider* the possibility that dead antelopes indicate danger. Although I concede the truth of (L), I maintain that the following psychological thesis is correct:

(P) $\Pr(O$ considers the possibility that $Z \,/\, O$ believes that X and believes that X supports $Y) > \Pr(O$ considers the possibility that $Z \,/\, O$ believes that $X)$.

(L) purports to describe a *logical* property of the confirmation relation; in contrast, (P) is a *psychological* claim that describes when individuals will consider certain hypotheses. My thesis is that thinking about tigers makes it more probable that the organism will consider the possibility that the dead antelope indicates danger. I am talking about the *context of discovery,* not the *context of justification.*

Proposition (P) is an empirical claim. If the problem about abandoning solipsism and the problem about using theories are structurally the same, it isn't that hard to test (P) by considering examples from the practice of science. We need to consider cases in which two groups of scientists confront a novel observation. For each, the problem is to say what predictive significance this new observation has. The two groups should be alike in terms of their antecedent observational knowledge. However, they differ in that one group has formulated a theory that they know is well-supported, whereas the other group has formulated no theory at all about the old batch of observations. Let us call the first group the Theoreticians and the second the Empiricists.

My claim is that the Theoreticians are more likely to solve the prediction problem than the Empiricists are. This is not to say that what the Theoreticians believe justifies a conclusion that is not justified by what the Empiricists believe. I am not contesting the truth of (L). Rather, I am advancing a thesis about what the two groups are likely to *consider* – (P) is the heart of the matter.

6. CONCLUDING REMARKS

A solipsistic organism is like a scientist without a theory. The organism can detect regularities in the course of its experience, but it will probably have difficulty interpreting experiences that are phenomenologically *new.* Rather, new experiences are apt to be handled by recording them as they occur; if they recur sufficiently often, the organism can gradually detect patterns that include the new type of experience. In short, new experiences are likely to start making sense only when they stop being new.

A nonsolipsistic organism, in contrast, is like a scientist with a theory. An experience that is phenomenologically novel can sometimes be interpreted in the preexisting theoretical framework. It isn't that the nonsolipsist that first spots a dead antelope is, then and there, absolutely certain that a tiger is present. Rather, this possibility is more likely to be considered. The nonsolipsist can drastically reduce the space of possible interpretations; a dead antelope is more likely to mean that a tiger is nigh than it is to mean that fish and chips are about to appear. The tiger theory and the other theories that the nonsolipsist has at hand shape the organism's interpretation of experience and direct it to ask some questions rather than others. Of course, none of this would do any good if these real-world theories were wildly misleading. It's a good idea to take seriously the possibility that tiger images and dead antelope images should be subsumed under one's theory of tigers only to the extent that dead antelopes are, in fact, a sign of tigers. A completely false theory can be worse than no theory at all.[14]

The basic picture I have tried to develop is that solipsists are limited to inductive inferences concerning regularities in the flow of experience whereas nonsolipsists can make abductive inferences concerning what lies outside the flow of experience. I have argued that the latter procedure is more powerful, not just in the obvious sense that it will lead to conclusions that are beyond the solipsist's ken, but in the sense that it is a more powerful guide to the course of experience itself. This enhanced power becomes visible when we consider the *changes* that occur in experience. It is the fact that *new* experiences occur, relative to the old ones already in place, that makes nonsolipsism useful.

Hypotheses of adaptive significance become testable when they predict variation. When a biologist describes the advantage a fish receives from schooling, the question arises as to why many fish do not school. A properly formulated adaptationist model should describe when the trait will be advantageous *and when it will fail to be.*

The organism I have described has a sensory system that is sufficiently developed that phenomenologically *new* sensations can arise during its lifetime. If an organism is sensitive just to light intensity, or to the acidity of its liquid medium, it is hard to see why positing a world outside of sensation would have a point. A second feature of my model is that the organism is not preprogrammed; it is not innately disposed to respond to novel experiences of the type under consideration. Many organisms have innate mechanisms of just this sort. If behavioral responses are entirely dictated by innate predispositions, my model predicts no advantage for the nonsolipsistic configuration.

In the descent of mind, it is not implausible to think that sensation predated cognition. Sensations make no reference to a world outside the organism; indeed, they make no reference, period. If an organism acquires the ability to formulate beliefs, or proto-beliefs, about its own sensory states, that organism is, so far, a solipsist. An organism of this type is subject to conditioning, which can cause it to associate one type of experience with another. The associations or "expectations" formed in this process are the ancestral precursors of predictions, and they are formulable within the confines of solipsism. Later on in evolution, organisms managed to break free of these confines. A more extensive type of cognition developed in which nonsolipsistic concepts were deployed. Rather than responding just to sensory states, the organism then exhibited the capacity to form representations that were stable over large variations in its sensations. Tiger images look different from dead antelope images, but the organism I have discussed manages to have a representation in the one case that is the same as a representation generated in the other. The utility of such common elements provided a reason to leave solipsism behind.

NOTES

1. See Essay 8 for further discussion of this inferential principle.
2. Putnam (1981), McDowell (1986), and Burge (1986) have each constructed semantic arguments against solipsism that depend on causal theories of representation in which the semantic contents of beliefs are individuated non-narrowly; Weitzman (unpublished) critically reviews these arguments. These arguments, even if correct, would not show that thoughts cannot be purely solipsistic. They endeavor to show that for a thought to have meaning, there must *exist* something outside the thinker's mind. However, it does not follow from these arguments that the thinker must form *beliefs* whose constituents purport to refer to these further items.

   A related set of questions is raised by Wittgenstein's private language argument. One aim of that argument was to demonstrate that (i) a meaningful term must have a publicly accessible "criterion" for correct application. However, even if (i) were correct, this would not entail that my meaningful terms cannot refer just to my own sensations. To reach that stronger conclusion, it also needs to be shown that (ii) a term that purports to refer just to one's own mental states doesn't have a public criterion. My hunch is that once "criterion" is clarified so as to make (i) plausible, (ii) turns out to be quite implausible.
3. In what follows, I won't address the question of why intentionality is advantageous. Rather, I will *assume* that the organisms in question have beliefs about something; then I'll inquire as to the utility of having nonsolipsistic rather than solipsistic beliefs.
4. Of course, if human thought and language are irreducibly committed to a nonsolipsistic framework, it will be impossible for us to formulate *in English* (or in any extant natural language) what the belief contents are of a solipsistic organism. If so, the examples just cited in the text should be viewed as approximations, which, at best, give a flavor of what solipsism is like. I take it that even if we are unable to formulate solipsistic beliefs, this does not show that such belief contents (or belief* contents*) are impossible for an organism that is differently constituted.
5. There is a related question about *causality.* Some disjunctive properties seem to be causally efficacious in spite of the fact that some of their disjuncts are not instantiated; we say that the toy was crushed because it was *under the tree* when the tree fell. But being under the tree is a disjunction

– it means being *somewhere* under the tree. However, other disjunctive properties seem to be causally inert, in spite of the fact that one of their disjuncts is causally efficacious. If a sack of oranges' *weighing five kilos* causes the spring balance to point to "11," it is decidedly odd to say that its having the property of *weighing five kilos or being bitter* caused the scale to do this. For further discussion of this difference, see Sober (1984, pp. 88–96).

6. My assumption that verificationism and its ilk are mistaken would be question-begging, if my goal were to address the epistemic question (E). But it is not. My target is the functional question (F). Even if the falsity of verificationism were enough to undermine solipsism, this would not explain what advantage there is in thinking in nonsolipsistic terms.
7. For further discussion of Dretske's theory, see Segal and Sober (1991).
8. This point about innateness, I believe, answers a question posed by Mark Sainsbury. Why can't the solipsist assign to states *A* and *B* the content "a dangerous experience is occurring"? I do not rule out the possibility that an organism might be innately programmed to form this content in the two circumstances. However, for a solipsist that learns from experience, I think it is improbable that this content will be triggered the first time it sees a dead antelope. The solipsist is less likely to consider the possibility that this novel experience is similar to other experiences that earlier proved to be signs of impending danger.
9. My answer to the functional question about operationalism does not entail that the meanings of theoretical terms never change. For all I've said, "mass" means one thing in Newtonian physics and something quite different in relativity theory. Substantive theoretical developments can motivate a change in the semantic properties of a term. The point is that we do not change the meaning we assign to a theoretical term just because a new measuring procedure is developed.
10. Although the present example concerns theory *augmentation,* parallel remarks apply to the issue of theory *correction.*
11. Although the Craigian replacements of the theories I have considered are quite simple, this isn't always so. When the Craigian replacements are more complicated than the theories they replace, the advantage of *coding economy* will be worth considering as an additional benefit of using the full-blown theory. It is well to remember, however, that coding economy is a concept that is language relative.
12. For related ideas on the import of prediction for other goals in science, see Forster and Sober (in press).
13. Ramsey (1931) discussed the idea that theoretical terms can be eliminated without impairing a theory's predictive power. His point was not to advocate the elimination of theoretical terms, but to raise the question of what contribution theoretical terms actually make. Referring to the definitions that would allow theoretical terms to be eliminated, he says (p. 230) that "rather than give all these definitions it would be simpler to leave the facts,

laws and consequences in the language of the primary system. Also the arbitrariness of the definitions makes it impossible for them to be adequate to the theory as something in process of growth." I believe that my argument in the text illustrates Ramsey's point about the import of diachronic considerations.

Commentators (e.g., Earman [1978]) on Ramsey usually emphasize the fact that the Craigian or Ramsified counterpart of a conjunction is not always equivalent to the conjunction of the counterparts: the constraint $C(X\&Y) = C(X)\&C(Y)$ is not always satisfied (though it happens to be satisfied in the example of T_1 and T_2 given in the text). The diachronic significance of this fact is that if one believes X and then learns Y, one may end up with a larger stock of empirical consequences than will be true if one believes $C(X)$ and then learns $C(Y)$.

Earman (1978) provides useful clarification of Ramsey's argument, which he construes as an argument for scientific realism. I do not interpret the point in this way, in that I see a rather sharp separation between the epistemic and the functional question. Even if the conjunction $X\&Y$ has more empirical consequences than the conjunction $C(X)\&C(Y)$, this is not, *per se,* a reason to believe $X\&Y$. See Forster and Sober (1994) for further discussion.

It is worth emphasizing that the fact that it makes sense to use theoretical terms leaves open what epistemic attitudes one should take to theoretical statements. Fictionalism and empiricism (e.g., of the sort defended by Van Fraassen [1980] and discussed in Essay 6 of the present volume) use theories, but decline to say that theories are true. The functional argument I am considering here concerns whether or not theories should be *eliminated,* not how a theory, once retained, should be *interpreted.*

14. Another assumption of my argument is worth registering here. The nonsolipsist performs computations that the solipsist does not execute. I assume that the energetic costs of these computations are low enough that they may safely be ignored.

REFERENCES

Burge, T. (1986). Cartesian error and the objectivity of perception. In P. Pettit and J. McDowell (eds.), *Subject, Thought, and Context.* Clarendon Press.

Dennett, D. (1978). *Brainstorms.* MIT Press.

Dretske, F. (1986). Misrepresentation. In R. Bogdan (ed.), *Belief.* Oxford University Press.

Dretske, F. (1988). *Explaining Behavior.* MIT Press.

Earman, J. (1978). Fairy tales vs. an ongoing story – Ramsey's neglected argument for scientific realism. *Philosophical Studies* 33: 195–202.

Forster, M., and Sober, E. (in press). How to tell when simpler, more unified, or less *ad hoc* theories will provide more accurate predictions. *British Journal for the Philosophy of Science.*

Gould, S. (1980). Natural selection and the human brain – Darwin versus Wallace. In *The Panda's Thumb.* Norton.

Hempel, C. (1965). The theoretician's dilemma. In *Aspects of Scientific Explanation and Other Essays.* Free Press.

McDowell, J. (1986). Singular thought and the extent of inner space. In P. Pettit and J. McDowell (eds.), *Subject, Thought, and Context.* Clarendon Press.

Orzack, S., and Sober, E. (1994). Optimality models and the test of adaptationism. *American Naturalist 143*:361–80.

Putnam, H. (1981). Brains in a vat. In *Reason, Truth, and History.* Cambridge University Press.

Ramsey, F. (1931). Theories. In *The Foundations of Mathematics.* Routledge and Kegan Paul.

Reichenbach, H. (1956). *The Direction of Time.* University of California Press.

Salmon, W. (1984). *Scientific Explanation and the Causal Structure of the World.* Princeton University Press.

Segal, G., and Sober, E. (1991). The causal efficacy of content. *Philosophical Studies 62:* 155–84.

Sober, E. (1984). *The Nature of Selection.* MIT Press.

Sober, E. (1988). *Reconstructing the Past: Parsimony, Evolution, and Inference.* MIT Press.

Sober, E. (1993). *Philosophy of Biology.* Westview Press.

Van Fraassen, B. (1980). *The Scientific Image.* Oxford University Press.

Weitzman, L. (unpublished). Skepticism and representation.

3

The adaptive advantage of learning and *a priori* prejudice

1. INTRODUCTION

Suppose an organism sees that a tiger is at hand. The organism must decide whether *this tiger is dangerous.* Two types of strategy are available for making this decision.

An individual who *learns* will believe the proposition, or believe its negation, depending on the character of its experiences. For example, the organism might attend to whether the tiger is wagging its tail, and decide what to believe on that basis. The alternative strategy is for the organism to decide on the basis of *a priori prejudice.* It will believe that the tiger is dangerous (or that it is not) irrespective of the character of its experience.

Learning is a conditional strategy, whereas adhering to an *a priori* prejudice is unconditional. The prejudiced individual conforms to the rule *always believe that tigers are dangerous.*[1] The learner conforms to the rule *believe that a tiger is dangerous if your experience has characteristic C, but believe that the tiger is not dangerous if your experience has characteristic D.*

The problem to be investigated here concerns the adaptive advantage of learning and *a priori* prejudice. Under what circumstances is learning advantageous? This problem is subsumable under the more general heading of determining when an *obligate response* is fitter than a *facultative response.* A polar bear can either have thick fur regardless of what the ambient temperature is, or it can have thick fur in some circum-

My thanks to Martin Barrett, Marc Bekoff, James Crow, Ellery Eells, Malcolm Forster, David Hull, Richard Lewontin, Greg Mougin, Michael Ruse, Alan Sidelle, Peter Godfrey Smith, Leora Weitzman, David S. Wilson, and Keith Yandell for their helpful suggestions.

stances and thin fur in others. Which strategy would be better for the polar bear to follow?

When I speak of an organism "deciding *a priori*" to believe that a tiger is dangerous, I do not mean that the organism is making a decision on its own. Rather, evolution has wired the organism to believe this proposition unconditionally. For exploratory purposes, I will adopt the working hypothesis that the traits in question are shaped by natural selection. So our question really is: When will natural selection favor learning over *a priori* prejudice?

In seeking an answer, we should not assume that the best policy with respect to one proposition will automatically be the best policy with respect to another. Maybe some problems are best approached by learning, while others are best solved *a priori.* And perhaps some organisms approach a given proposition via learning while others settle the same matter *a priori.* What is an *a priori* prejudice for one species may be a product of learning for another.

2. THE MODEL

Whether learning is better than prejudice depends on a variety of factors, which we now must try to identify. We begin by specifying the costs and benefits imposed by true and false belief:

	A is true	*A* is false
Believe *A*	$x + b_1$	$x - c_2$
Believe *not-A*	$x - c_1$	$x + b_2$

It is not an assumption of this model that true belief is superior to falsehood. In what follows we will consider this idea (that $x + b_i > x - c_i$, $i = 1,2$) as one possibility among several.[2]

We also must consider the probability that the agent who learns will form a given belief, conditional on the way the world is:

	A is true	*A* is false
Believe *A*	$1 - a$	n
Believe *not-A*	a	$1 - n$

In this model, *a* and *n* are *error probabilities;* low values for *a* and *n* mean that if *A* is true (false), the agent will probably come to believe that *A* is true (false).

I will depart somewhat from scientific usage by saying that *a* and *n* measure how *sensitive* the organism is. They describe how probable it is that the organism will believe a given proposition, if that proposition is true. Sensitivity, so defined, is a quite different matter from *reliability.* Reliability means that a proposition is probably true, if the agent believes that it is. As I use the terms, sensitivity is a world-to-head relation, while reliability is a head-to-world relation.[3] It is not hard to see how the concepts differ. An agent who believes only those propositions that are absolutely certain will be quite reliable, though very insensitive.[4] In any event, our model, as described so far, describes sensitivity, not reliability, though we will come to reliability in a while.

We now can describe three strategies. First, there are two forms of prejudice to consider. An organism can unconditionally believe that A is true or unconditionally believe that A is false. The third strategy is to decide whether to believe A or *not-A* based on the learning mechanism whose error characteristics were just described. Where p is the probability that A is true,[5] the fitnesses of the three strategies are

$$\begin{aligned} w_{\text{Prej}(A)} &= (x + b_1)p + (x - c_2)(1 - p) \\ w_{\text{Prej(not-}A)} &= (x - c_1)p + (x + b_2)(1 - p) \\ w_{\text{Learn}} &= [(x + b_1)(1 - a) + (x - c_1)a]p \\ &\quad + [(x + b_2)(1 - n) + (x - c_2)n](1 - p) \end{aligned}$$

Two inequalities follow:

(1) $w_{\text{Learn}} > w_{\text{Prej}(A)}$ if and only if
$$(b_2 + c_2)(1 - n)(1 - p) > (b_i + c_1)ap.$$

(2) $w_{\text{Learn}} > w_{\text{Prej}(not\text{-}A)}$ if and only if
$$(b_1 + c_1)p(1 - a) > (b_2 + c_2)(1 - p)n.$$

3. A PROPOSITION'S IMPORTANCE

Exploring the implications of this model requires that we define a new concept:

$$\text{the importance of a proposition } A_i =_{\text{def}} b_i + c_i.$$

The importance of a proposition A_i is measured by seeing *how much difference* it makes whether one believes A_i or believes *not*-A_i, if A_i is true; $(x + b_i) - (x - c_i) = b_i + c_i$.

This definition induces a threefold division among propositions: a proposition A_i can have positive importance ($b_i + c_i > 0$), negative importance ($b_i + c_i < 0$), or zero importance ($b_i + c_i = 0$). Examples of the first type are familiar: If a mushroom is poisonous, it may be better to believe this than to believe its negation. For examples of the third kind, the reader need look no farther than the propositions of the present essay. I suspect that they are of no importance; even if they are true, it makes no difference in fitness whether you believe them or their negations.

The second sort of proposition is the least familiar of the three, but it is by no means impossible. A contemporary of Darwin's once opined that it would be better not to believe the theory of evolution even if it happens to be true (Ruse 1979). The same has been said more recently of utilitarianism (Williams 1972). And on a more personal level, the thought has been formulated that we should believe that life has a meaning, even if it does not. Of course, I am not endorsing any of these claims; I mention them just to illustrate what it means to say that a proposition has negative importance.

It is plausible that some propositions and their negations will both have positive importance. For example, in the pair of propositions (*A tiger is nearby, No tiger is nearby*), it may be better to believe the truth, whatever that happens to be. But in other examples, a different pattern may obtain. For example, in "The Will to Believe" William James suggests that it is better to believe in God, regardless of whether God in fact exists. One receives comfort either way, and that is the main thing (James 1897). This amounts to saying that the proposition "God exists" has positive importance and that "God does not exist" has negative importance. Again my point is not to endorse the claim, but to illustrate how a proposition and its negation need not have the same type of importance.

4. SOME CONSEQUENCES

We may begin with the question of how the relationship of learning to *a priori* prejudice is affected by the issue of whether true beliefs are better than false ones. As noted earlier, there is no requirement that this be so, but a weaker requirement is nonetheless entailed by (1) and (2):

(3) If $w_{\text{Learn}} > w_{\text{Prej}(A)}, w_{\text{Prej}(not\text{-}A)}$, then $(b_1 + c_1)$ and $(b_2 + c_2)$ must differ from 0 and must have the same sign.

If a proposition or its negation is of no importance ($b_i + c_i = 0$, for some value of i), then learning cannot be superior to both forms of prejudice. And if $(b_1 + c_1)$ is positive and $(b_2 + c_2)$ is negative, then it is better to believe proposition *A whether or not A is true.* When the payoffs are structured in this way, selection will favor some form of *a priori* prejudice over learning.

Proposition (3) describes two circumstances in which learning may be superior to *a priori* prejudice. In the first, true belief is *always* more advantageous than false belief. In the other, false belief is *always* better than truth. But propositions that don't matter, and ones that you are better off believing regardless of whether they are true, are ruled out.

Strictly as a matter of convenience, I will assume in (most of) what follows that true belief is better than false belief. This assumption is not substantive, since counterparts of the results I will derive could be obtained under the opposite assumption.

Given the assumption that $(b_1 + c_1)$ and $(b_2 + c_2)$ are both positive, we can obtain from (1) and (2) a necessary and sufficient condition for learning to be better than both forms of *a priori* prejudice:

(4) If $(b_1 + c_1)$ and $(b_2 + c_2)$ are both positive, then $w_{\text{Learn}} > w_{\text{Prej(A)}}, w_{\text{Prej}(not\text{-}A)}$ if and only if
$p(1 - a)/[(1 - p)n] > (b_2 + c_2)/(b_1 + c_1) > ap/[(1 - n)(1 - p)]$.[6]

From the criterion provided by proposition (4), we can extract a result concerning the values of the error probabilities a and n associated with the learning mechanism. Notice that it follows from (4) that

If $(b_1 + c_1)$ and $(b_2 + c_2)$ are both positive, then $w_{\text{Learn}} > w_{\text{Prej}(A)}, w_{\text{Prej}(not\text{-}A)}$ only if

$$p(1 - a)/[(1 - p)n] > ap/[(1 - n)(1 - p)],$$

which simplifies to

(5) If $(b_1 + c_1)$ and $(b_2 + c_2)$ are both positive, then $w_{\text{Learn}} > w_{\text{Prej}(A)}, w_{\text{Prej}(not\text{-}A)}$ only if $1 > a + n$.

A necessary condition for learning to outperform prejudice is that the sum of the error probabilities associated with learning be less than unity. Learning demands at least this minimal degree of sensitivity. Proposition (5) says that learning is superior to prejudice only if believing a proposition is *correlated* with the proposition's being true:

$$\Pr(\text{Believe } A_i / A_i) > \Pr(\text{Believe } A_i / not\text{-}A_i), \qquad i = 1,2.$$ [7]

It is interesting that the necessary condition described by (5) is not sufficient for learning to be fitter than *a priori* prejudice. Even when the learning device effects a correlation between belief and truth, it still may not make sense to deploy the learning device.[8]

In addition to the requirements given by propositions (3) and (5), a third necessary condition must be satisfied, if learning is to be fitter than *a priori* prejudice:

(6) $w_{\text{Learn}} > w_{\text{Prej}(A)}, w_{\text{Prej}(not\text{-}A)}$ only if $p \neq 0,1$.

There is no advantage to deploying a learning device to reach a verdict on propositions that *must* be true, or that have no chance of being true.

The concept of a proposition's importance allows us to deduce which *a priori* prejudice is fitter:

(7) $w_{\text{Prej}(A)} > w_{\text{Prej}(not\text{-}A)}$ if and only if $p(b_1 + c_1) > (1 - p)(b_2 + c_2)$.

If one is going to choose an *a priori* prejudice, the better choice is the proposition with the greater *expected* importance. Notice that even if proposition A cannot fail to be true, it is not inevitable that it should be an *a priori* prejudice. There is no advantage in having as an *a priori* prejudice a proposition that is of no importance, even if that proposition must be true.

Combining propositions (6) and (7), we may conclude that necessary truths that have some degree of positive importance will be *a priori.*[9] In contrast, it is not inevitable that *a priori* beliefs should be necessary.

We next turn to the issue of reliability. Proposition (1) entails the following:

If $w_{\text{Learn}} > w_{\text{Prej}(A)}$ and A is more important than *not-A*, then
$(1 - n)(1 - p) > ap$.

Notice that

$(1 - n)(1 - p) > ap$ if and only if
Pr(*not-A* & Believe *not-A*) > Pr(*A* & Believe *not-A*) if and only if
Pr(*not-A* / Believe *not-A*)Pr(Believe *not-A*) >
Pr(*A* / Believe *not-A*)Pr(Believe *not-A*).

This means that

(8a) If $w_{\text{Learn}} > w_{\text{Prej}(A)}$ and A is more important than *not-A*, then
Pr(*not-A* / Believe *not-A*) > .5.

That is, if learning is fitter than prejudice, then the learner's reliability with respect to the *less* important proposition must exceed the threshold specified. And if the two propositions are equally important, the learner must be more than minimally reliable with respect to both, if learning is to be better than prejudice.

Proposition (8a) is interesting because we began our discussion of learning by focusing on the issue of sensitivity, which was defined as a world-to-head relation. For learning to be superior to prejudice, we now find that a head-to-world relation also must obtain.

This head-to-world conclusion is relevant to the subject of philosophical skepticism. Skepticism comes in many forms; the one that is pertinent here inquires why we should think that the beliefs an organism forms on the basis of sensory evidence are at all reliable. When an organism looks out into a field and thinks "there is a tiger," what is the probability that what the organism thinks is true? The Cartesian skeptic asks us to answer this question by appeal only to propositions that are absolutely certain. But in a more naturalistic vein, we might consider whether evolutionary considerations entail that the organism's belief is probably correct. The hope for a result of this sort may seem dim, given that Mother Nature usually goes in for comparative, not absolute, judgments; the criterion for a trait to evolve under natural selection is that it be fitter than the competition;[10] normally, it need not satisfy any absolute requirement of adaptedness. This may encourage us to imagine that there is no limit to how *un*reliable a perceptual device can be and still evolve.

Proposition (8a) shows that this pessimism is exaggerated. Consider an organism that decides whether a tiger is present by consulting the deliverances of a learning device. It is often plausible to assume that for this strategy to have evolved, it must have done better than either form of *a priori* prejudice. This doesn't quite entail that the organism's positive and negative tiger beliefs are probably correct. What we find is something more modest, though it is interesting nonetheless: The less important proposition in the pair (*a tiger is present, no tiger is present*) is probably true each time it is believed.[11]

We have just seen that the model under consideration says that when learning is fitter than prejudice, learners must be more than minimally reliable. In fact, proposition (8a) slightly understates the relevant results. No matter how propositions *A* and *not-A* compare in terms of their relative importance, some consequence about the reliability of learners is entailed. A fuller statement of the idea is

(8b) If $w_{\text{Learn}} > w_{\text{Prej}(A)}, w_{\text{Prej}(not\text{-}A)}$, then:
Pr(*not-A* / Believe *not-A*) > .5 if
A is more important than *not-A*,
Pr(*A* / Believe *A*) > .5 if *A* is less important than *not-A*, and
Pr(*not-A* / Believe *not-A*) > .5 and Pr(*A* / Believe *A*) > .5 if
A and *not-A* are equally important.

Suppose that learning evolves to fixation (100 percent representation in the population) because it is fitter than both forms of *a priori* prejudice.

Proposition (8b) says that if we look at the population after learning has gone to fixation, we will find that the organisms are reliable with respect to one or the other or both of the propositions (*A* and *not-A*) that they find themselves believing.

Now let us consider the symmetric question: If some form of *a priori* prejudice is fitter than learning, does it follow that the prejudiced belief will probably be true? Interestingly, the answer is *no.* Let us consider a case in which $w_{\text{Prej}(A)} > w_{\text{Prej}(not\text{-}A)} > w_{\text{Learn}}$. Learning is least fit of all, we will suppose, because the error probabilities (a and n) of the learning device are large. And proposition (7) specifies a criterion for the first prejudice to be fitter than the second; if $w_{\text{Prej}(A)} > w_{\text{Prej}(not\text{-}A)}$, then $p(b_1 + c_1) > (1 - p)(b_2 + c_2)$. The point to notice so far is that this last inequality places no lower bound on the value of p.

Someone with the *a priori* prejudice that proposition *A* is true will *always* believe that proposition. For such individuals, Pr(*A* is true / Believe that *A*) = p. So even if believing *A* as a matter of *a priori* prejudice is the fittest of the three strategies, nothing follows about the probability that this belief will be true.

To be sure, we can obtain a somewhat limited result concerning the reliability of *a priori* prejudice. Proposition (7) tells us straightforwardly that

(8c) If $w_{\text{Prej}(A)} > w_{\text{Prej}(not\text{-}A)}, w_{\text{Learn}}$ and *A* is not more important than *not-A,* then $p > .5$.

However, the point is that if *A is* the more important proposition and Prej(*A*) is the fittest of the three strategies, nothing follows about the reliability of the prejudice that evolves.

This asymmetry between (8b) and (8c) may have some relevance to the perennial antagonism between empiricism and rationalism. Is experience or *a priori* reflection a more reliable guide to the way the world is? The present model inclines toward the empiricist answer. If learning and prejudice compete against each other and learning evolves, then the deliverances of the learning device will be more than minimally reliable. However, if the competition is resolved by having an *a priori* prejudice evolve, nothing follows concerning the prejudice's reliability. This does not mean that the prejudice must be *un*reliable. But it does suggest that if learning is to evolve, it must pass a more stringent test than the prejudices with which it competes.

5. DISCUSSION

It is important that propositions (5) and (8b) not be overinterpreted. They describe minimum levels of sensitivity and reliability that a learning device must satisfy if it is to be fitter than the strategy of *a priori* prejudice. This, by itself, does not guarantee that the learning devices used by extant organisms live up to these minimum standards. To begin with, it is not automatic that these learning devices evolved in a population in which they competed with *a priori* prejudice. In addition, evolutionary models describe ancestral conditions, which may not be the same as the conditions that are in place today. Even if our ancestors excelled at identifying tigers, this says nothing about how well we do in recognizing Toyotas. And finally, the model shows that learning will be fitter than *a priori* prejudice only if the truth value of a belief makes a difference to the survival and reproduction of the believer. Although this may be intuitive for the contrast between believing that *a tiger is present* and believing that *no tiger is present,* it is less than obvious for other pairs of contrasting alternatives.

The concept of learning addressed in this model is a very limited one. The model describes learners as developing beliefs in response to environmental cues; they do not have access to the successes and failures they encountered on previous occasions. The individuals described here learn from their *present* environment, not from their experiences in the *past.* Obviously a more complete exploration of the advantages of learning would have to take these diachronic issues into account.[12]

There is one more limitation of this model that I should mention. The model treats each proposition as a separate problem. It predicts how a given proposition will be handled by attending to the proposition's importance and to the sensitivity of the learning device at hand. But it is questionable, especially in the case of our own species, that each proposition involves an independent problem.

Here I am alluding to the *systematic* quality of our cognitive apparatus. Perhaps the fitness of our ancestors was affected by their ability to tell whether a tiger is nearby. And perhaps it made no difference whether they could tell whether tiger lilies are at hand. However, it does not follow from this that the two propositions will be handled differently. For it may be true that any learning device that excels on the first task will probably excel on the second. If so, we may possess learning devices that deliver verdicts on propositions of no importance.[13] This spin-off scenario for the evolution of cognitive capacities is important to consider, but it is not reflected in the present model.

With these *caveats* in mind, we can summarize what the model says by asking when *a priori* prejudice makes sense. The following three considerations help make it better for an organism to believe proposition *A a priori,* rather than to decide whether *A* is true by learning:

(i) The error probabilities (a and n) of the learning device are large.
(ii) *A* is probably true (p is large).
(iii) *A* is more important than *not-A* ($b_1 + c_1 > b_2 + c_2$).

Symmetrically, we can summarize the factors that help make learning a better strategy than either type of *a priori* prejudice:

(iv) The error probabilities (a and n) of the learning device are small.
(v) *A* and *not-A* have middling probabilities (p is neither large nor small).
(vi) *A* and *not-A* are about equally important ($b_1 + c_1 \approx b_2 + c_2$).

These factors are not separately necessary, but are considerations that promote the relevant fitness differences, all else being equal. Their interaction is depicted in Figure 3.1.

We can use these results to predict which propositions an organism will probably approach by learning, and which by *a priori* prejudice. If we assume that these strategies have been shaped by natural selection as modeled here, we can expect organisms to treat different propositions differently. We would not expect organisms to be pure *tabula rasas; a priori* prejudice will probably find some role in the organism's cognitive economy. Propositions that are important, hard to learn, and usually true can be expected to be furnished as *a priori* prejudices.

As a first approximation, the principle of empiricism can be taken to assert that sensory experience is the only source of information about the world.[14] Those attracted by empiricism are often stymied by the problem of how the principle can be defended while remaining true to empiricist tenets. Can an empirical defense of the principle be given that avoids begging the question? If the present model is any guide, this hard question need not detain us. The reason is that the theory of evolution provides an empirical reason to doubt that empiricism is correct.[15]

6. AGNOSTICISM

In the above model, the only epistemic attitudes the organism can take to a given proposition is believing it or believing its negation. However,

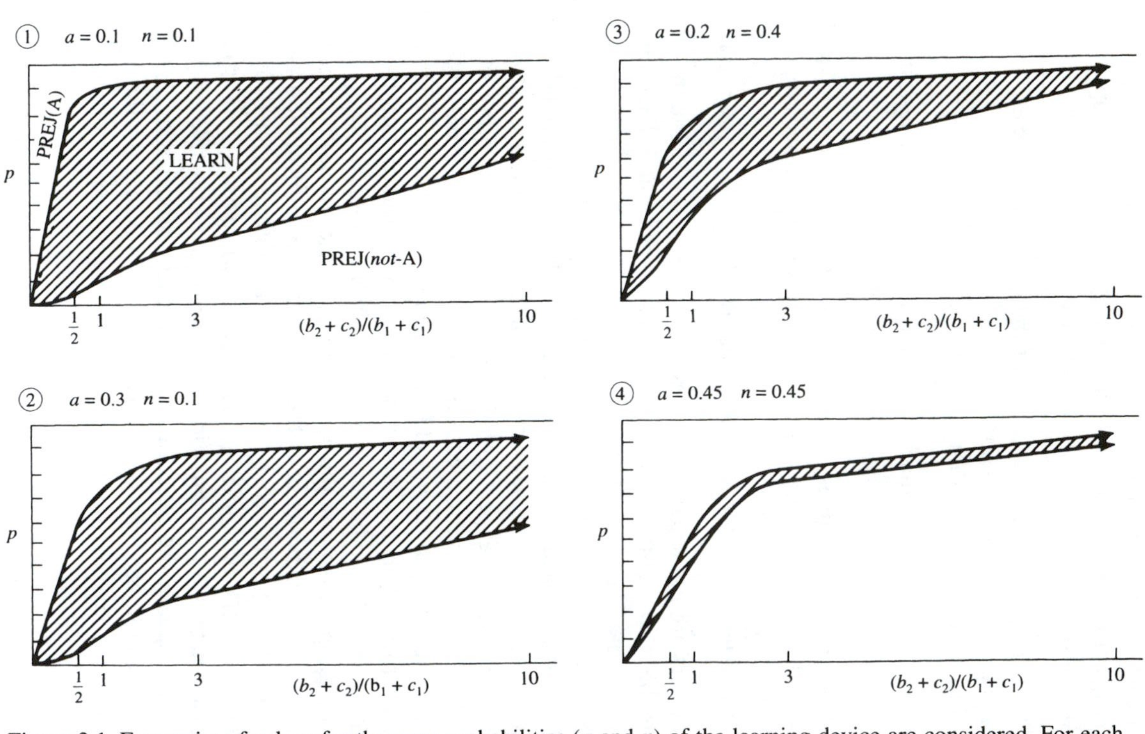

Figure 3.1 Four pairs of values for the error probabilities (a and n) of the learning device are considered. For each pair of values, learning is fitter than *a priori* prejudice within one region of parameter space, but the reverse is true outside that region.

a third option is worth considering; the organism can suspend judgment and be an agnostic.

The question to investigate is how this third option is related to the two already discussed. I will assume that the payoffs are structured as follows:

	A is true	*A* is false
Believe *A*	$x + b_1$	$y - c_2$
Agnosticism	x	y
Believe *not-A*	$x - c_1$	$y + b_2$

It is an open question whether these three epistemic attitudes are behaviorally different in a given setting. Maybe organisms who believe that a tiger is nearby behave differently from those who believe that none is present, but those with the latter belief behave exactly like organisms who have no beliefs on the matter, one way or the other. This could be accommodated in the above model by setting $c_1 = b_2 = 0$. Notice also that agnosticism may have different consequences for the organism, depending on whether *A* is true or false.

We now can calculate how agnosticism compares with the two forms of *a priori* prejudice:

(9) $w_{\text{Prej}(A)} > w_{\text{Agnostic}}$ if and only if $pb_1 > (1 - p)c_2$.
(10) $w_{\text{Prej}(not\text{-}A)} > w_{\text{Agnostic}}$ if and only if $(1 - p)b_2 > pc_1$.

Unsurprisingly, a prejudice is better than agnosticism if the expected benefit of the prejudice's being true exceeds the expected penalty of the prejudice's being false. Notice that it is possible for *both* forms of *a priori* prejudice to be superior to agnosticism. Simply let $b_1 >> c_2$ and $b_2 >> c_1$. Nonetheless, if *A* is vastly more probable than its negation, one might expect Prej(*A*), but not Prej(*not-A*), to be better than agnosticism.

The next issue concerns the relation of learning and agnosticism. The relevant proposition is

(11) $w_{\text{Learn}} > w_{\text{Agnostic}}$ if and only if
$$pb_1(1 - a) + (1 - p)b_2(1 - n) > pc_1a + (1 - p)c_2n$$

Just as in propositions (9) and (10), belief based on learning is superior to agnosticism precisely when the expected benefit of true belief exceeds the expected cost of false belief.[16]

It is worth pointing out that the present treatment of agnosticism does not involve changing how the learning strategy was described earlier. Learners believe or disbelieve the proposition at issue; they do not suspend judgment. A more sophisticated conception of learning would include this third option.[17] However, a model of this type is beyond the scope of the present paper.

7. INNATENESS AND A PRIORICITY

What does it mean for a proposition to be *a priori?* A standard definition is that the proposition can be learned or justified by reason alone, once the constituent concepts are in hand. To say that *bachelors are unmarried* is an *a priori* proposition doesn't mean that people are born knowing that this proposition is true. It may be that the concept of bachelorhood has to be acquired. In addition, *a prioricity* has to do with how a proposition *could* be learned or justified, not with how it was learned or justified in fact. For example, the nineteenth-century mathematician Plateau solved the problem of determining the surface of least area that fills a given contour by dipping a wire frame into a tub of soap suds (Courant and Robbins 1969). In some sense, Plateau *could* have figured this out by reason alone and so the proposition in question is *a priori;* yet his *de facto* method was empirical.

Although some philosophers maintain that all analytic truths are innate,[18] I see no reason to follow them here. Perhaps many *a priori* truths are acquired by social learning. For example, maybe almost all people who believe the prime number theorem (that there is no largest prime) come to this belief by learning it from others. And maybe people who believe that all bachelors are unmarried do so because of experiences they have had in language learning.

So the real subject of this paper is innateness, not *a prioricity.* However, a word of clarification is needed in connection with the concept of innateness.

Some songbirds produce the song appropriate for their species only when they hear it first. Other species of songbird are able to produce their characteristic song even when reared in silence. And a third type of songbird will produce the song unique to its species only when it hears some song or other; even when it hears a quite different song from a different species, this will trigger the song of its own species (Gould and Marler 1991).

Should we apply the concept of "innateness" to this third case? Is the song learned from experience or is the song "preformed" inside the bird, only awaiting an environmental trigger to emerge?[19] I am skeptical that this preformationist alternative makes literal sense. Surely it makes no sense for many nonmental traits. For example, human beings grow pubic hair when they reach adolescence, but it would be absurd to say that pubic hair is "in" human babies from the time they are born.

In any event, the model provided in this paper does not require that behaviors be dichotomized in this way. We needn't decide whether to classify the bird song as learned or innate. Rather, what is important is that we can say which environmental conditions a behavior or a belief depends upon. A phenotype can be more or less conditional on the state of the environment. It may emerge whether or not the bird hears anything at all. Or it may require hearing some sort of bird song. Or it may require hearing a particular sequence of notes. And so on.[20]

The fundamental problem addressed here is whether a belief (or, more generally, a phenotype) should be conditional on some environmental cue. For the sake of convenience, I have classified beliefs that are relatively invariant over the organism's experiences as "*a priori* prejudices" and ones that depend on the details of experience as "learned." However, the underlying reality is a continuum.

8. WHEN WILL KNOWING EVOLVE?

Let us say that an organism knows that proposition A is true precisely when the organism truly believes the proposition and Pr(A is true / Believes that A) = 1.0. Knowledge, so defined, does not require the ability to formulate a proof that A is true, nor does it require that A be introspectively beyond doubt.[21] Rather, knowledge is just perfectly reliable true belief.

It is quite clear from the preceding model that belief fixation devices, whether they involve learning or *a priori* prejudice, need not provide knowledge if they are to evolve.[22] For a strategy to evolve, there is no requirement that Pr(A is true / Believes that A) = 1. Still, we can investigate what must be true for knowledge to arise as a *by-product* of selection. What must be true if an evolved belief fixation device is to give us knowledge?

We can divide this question in two. First, let us suppose that learning is fitter than *a priori* prejudice as a means to deciding whether proposition A is true. This means that learning will evolve to fixation. We now can use Bayes' theorem to describe what knowing requires for the resulting organisms:

Pr(A is true / Believes that A) =
Pr(Believes that A / A is true)Pr(A is true) / Pr(Believes that A) =
$(1 - a)p / [(1 - a)p + n(1 - p)]$, if the individual is a learner.

It follows that

(12) Learners know that A when they believe A if and only if $n = 0$ or $p = 1$.

We can ask the same question about the *a priori* prejudice that A is true. Let us suppose that this *a priori* prejudice is the fittest strategy and that it therefore evolves to fixation. If we examine an organism in the resulting population and ask whether its belief that A counts as knowledge, we again can use Bayes' theorem to obtain an answer. For convenience, I will represent the properties of the prejudiced individual in terms of the error probability a:

Pr(A is true / Believes that A) = $(1 - a)p$, if the individual believes A as an *a priori* prejudice.

If follows that

(13) Prejudiced individuals know that A when they believe A if and only if $a = 0$ and $p = 1$.

The contrast between propositions (12) and (13) is of some interest. The criterion for empirical knowledge is a disjunction, whereas the criterion for *a priori* knowledge is a conjunction. This suggests that *a priori* knowledge may be "harder" to come by than empirical knowledge, though the conditions on both are quite formidable. It also is worth noting that sensitivity to proposition A (represented by the value of a) figures as a necessary condition in the criterion for *a priori* knowledge, but plays no role at all in the criterion for empirical knowledge. Symmetrically, sensitivity to *not-A* (represented by the value of n) is relevant to the question of whether the organism has empirical knowledge that A is true, but is irrelevant to the question of whether the organism knows A as a matter of *a priori* prejudice.[23]

Propositions (12) and (13) make no mention of costs and benefits. These quantities are obviously important in determining which strategy evolves. But once a strategy has evolved, the question of whether it produces knowledge has nothing to do with what is good for the organism.

9. WHEN IS IT WORTH BEING MORE SENSITIVE?

The model described so far assesses the fitness of a strategy by mapping the costs and benefits of the beliefs that the strategy produces. However, the fitness of a piece of machinery – cognitive or otherwise – depends not only on its products, but on the "internal costs" of building and maintaining the machinery itself. For example, if an organism is going to learn whether proposition A is true, it must be equipped with sensors and with central processors that allow it to interpret the testimony of its senses. These can cost calories to build and maintain. By the same token, if an organism is going to believe proposition A as a matter of *a priori* prejudice, the organism will require machinery that allows it to do this, and this machinery will impose an internal cost as well. One should not assume that these costs are irrelevant to the problem of comparing the adaptive advantages of different strategies.

Taking account of internal costs is especially important when one considers different learning strategies. Suppose an old learning strategy has an internal cost of I_{old} while a new strategy has an internal cost of I_{new}. If a_{old}, n_{old}, a_{new}, and n_{new} are the error probabilities associated with the old and new mechanisms, respectively, we can say when the new mechanism is fitter than the old:

$$w_{\text{Learn-new}} > w_{\text{Learn-old}} \text{ if and only if}$$
$$I_{old} - I_{new} > p(a_{new} - a_{old})(b_1 + c_1) + (1 - p)(n_{new} - n_{old})(b_2 + c_2).$$

If the new mechanism has a higher internal cost than the old one, then at least one of the error probabilities of the new device must be lower than the counterpart error probability of the old device. The exact reduction that is required depends on the expected importance of the two propositions. This result underwrites the intuitive idea that greater precision in belief formation (or in phenotypic response generally) is sometimes not worth the energetic candle.[24] Improvements in fitness sometimes come by adding new equipment; at other times, the advantage goes to leaving costly equipment behind.

When learning and *a priori* prejudice are compared, it is tempting to assume that the former must impose a higher internal cost than the latter. Whether this intuition is sustainable as a generalization about cognitive adaptation is an interesting problem. However, it is not always true when a facultative adaptation is compared with one that is obligate. Consider an example of the so-called Baldwin effect.[25] Animals are often born with calluses on their palms and feet. Yet it is clearly possible to form calluses solely as a facultative response to friction.

Why are calluses given to us, so to speak, as *a priori* prejudices, rather than as something that we have to acquire?

If the trait to consider is a *healthy* (uninfected) callus, then the facultative acquisition of this trait has nontrivial error probabilities; there is a serious chance that friction to the skin will fail to produce a healthy callus. What is more, in most environments, an organism's palms and feet are abraded, and in such environments, it is *important* that the organism have calluses. Plausibly, the internal cost of making calluses innate is higher than the internal cost of having them be acquired by way of friction to the skin after the organism is born. So the trait became obligate, rather than facultative, even though the internal cost of the *a priori* prejudice is higher than the internal cost of the learning device.

10. CONCLUDING REMARKS

As noted at the outset in connection with the polar bear's fur, there is nothing especially cognitive about the model described here.[26] Polar bears can either have thick fur unconditionally, or they can vary the thickness of their fur in accordance with the deliverances of some measurement device[27] that has specifiable error characteristics. Belief never enters into the picture. It is arguable, though it is not obvious, that the polar bear's problem is fundamentally different from that of a cognitive agent. Perhaps beliefs are products of cognitive strategies in a way that fundamentally differs from the way that fur is a product of thermal adaptation. I see no reason to believe this, but it is certainly a problem worth exploring.

In assimilating the problem of learning and *a priori* prejudice to the polar bear's fur, I have set to one side the general problem of the adaptive significance of cognition. I have not asked: Given that a phenotype is to be facultative (or obligate), why should it be placed under the control of a cognitive mechanism? Phenotypes are often dependent on environmental conditions without mental representations entering the story. Why belief should play a role is an interesting question, but it is not the one I have addressed. Rather, I have *assumed* that the phenotype formed by a strategy is a belief; my question has been whether the formation of the belief should be conditional on the state of the environment.

It will be evident that I have approached the issue of learning and innateness from an angle that is somewhat nonstandard. Psychologists have considered behavioral evidence for and against hypotheses of innateness (e.g., via poverty of the stimulus arguments). Philosophers have been attracted by issues of conceptual clarification: What does it mean to say that an agent believes some proposition innately, and how is this different from saying that the individual has the innate capacity to acquire the belief? For these epistemic and semantic questions, I have substituted a question of functional utility: When is it advantageous to believe a proposition innately and when is it better to have belief depend on a learning process? This evolutionary question does not replace the two more traditional queries, but I believe it has sufficient interest to merit investigation on its own.

NOTES

1. The quantifier "always" ranges over possible experiences. Also, I say that organisms *conform* to rules, rather than *follow* rules, because I don't want to require that an organism cognitively represent the strategy it uses. In the same sense, planets conform to, but do not follow, the laws of motion.
2. For convenience, I will sometimes refer to the proposition that A is false as A_2.
3. I owe this head/world terminology to Godfrey Smith (1991), who credits it to Field (1990). The present essay has several points of contact with Godfrey Smith's, which provides a definition of "optimal signal detection."
4. More generally, note that Bayes' theorem entails that

   $$\Pr(\text{Believe}A\,/\,A) = \Pr(A\,/\,\text{Believe }A)\Pr(\text{Believe }A)\,/\,\Pr(A).$$

   The left-hand term describes the agent's sensitivity, while the conditional probability on the right describes the agent's reliability. This shows why high sensitivity does not entail high reliability, or conversely.
5. Just as would be the case in a model of the advantage of phenotypic plasticity, the probability p should be understood objectively, not as reflecting the degree of belief, if any, of the organism.
6. If $(b_1 + c_1)$ and $(b_2 + c_2)$ are both negative, then the inequalities in the criterion would both reverse.
7. The more general point, of which (5) is an instance, is that learning is better than prejudice only if the learning device associates *advantageous* belief states with conditions that obtain in the world. It is a separate matter whether the organism is better off believing A or not-A, when A is true.
8. And this for reasons quite separate from the matter of "internal costs," to be discussed in Section 9.
9. Of course, an organism can't believe a given proposition if it lacks the concepts out of which the proposition is constructed. So it is possible for

there to be important necessary propositions that an organism fails to believe, and so fails to have as *a priori* prejudices.

10. This is the standard assumption in phenotypic models of selection; it can be false if the genetic system "gets in the way." See Section 5.2 of Sober (1993) for discussion of this point.
11. We find here a pale reflection of Donald Davidson's (1984) claim that most of our beliefs must be true. Of course, the argument I have presented is entirely different from Davidson's; nor is the conclusion as general as his. However, it does show that, in expectation, most of the beliefs that fall in a specific class (i.e., ones that are less important than their negations and are acquired by a learning mechanism that evolved in the manner specified) are true. This issue is explored more fully in Essay 4.
12. See, for example, the learning strategies discussed in Pulliam and Dunford (1990).
13. It also is possible that different "modules" should deliver different verdicts on the very same proposition. A proposition *P* may be enshrined as an *a priori* prejudice and yet a learning device may tell the organism that *P* is false. Just as two environmental indicators may come into conflict, so an environmental indicator may conflict with an *a priori* prejudice.
14. I owe this informal rendition of the principle to Bas Van Fraassen. Notwithstanding the fact that the concept of "information" needs to be clarified, this statement of the principle is a good place to begin.
15. In spite of this negative verdict, I believe that there are important empiricist ideas that ought to be preserved in an adequate epistemology. This point is elaborated in Essay 6.
16. In "The Will to Believe," William James (1897) remarks that "he who says 'Better go without belief forever than believe a lie!' merely shows his own preponderant private horror of becoming a dupe." In light of the above model, it is no surprise that Mother Nature did not equip us with the Cartesian policy of remaining agnostic toward all propositions except those that are subjectively (or objectively) certain.
17. Indeed, it may be better still to move to a degree of belief conception, in which there are far more than three epistemic attitudes to consider. In doing so, the problem of how to assign a degree of belief to a proposition will be separated from the problem of how to behave, since action will be a joint function of probabilities and utilities (and a decision rule). The present model, in which the agent either believes or disbelieves a proposition, posits too close a connection between belief and action.

    A further advantage of the degree of belief approach is that the model of phenotypic plasticity described by D. W. Stephens (1989) can be applied to the case of belief. Stephens' approach, which is based on the work of the economist J. P. Gould (1974), requires that the phenotypes considered be located on a metric. Although it is difficult to see how to do this for a set of mutually exclusive beliefs, there is no such problem when one considers different degrees of belief that an agent might have with respect to a single proposition.

It is worth mentioning that the Stephens–Gould definition of the "value of information" entails that learning is never worse than *a priori* prejudice. Of course, no such result follows from the framework developed here. This clash is only apparent; it arises from the way that Stephens and Gould define the concept of a learning strategy.

18. This is the position taken by Fodor (1975). At most, Fodor's argument shows that concepts cannot be *learned.* But that isn't enough to show that they are innate. When I go to Florida, I acquire a suntan, but not by learning.
19. Fodor (1981) argues that the difference between learning and triggering is fundamental to the problem of innateness.
20. If talk of innatism sometimes carries with it an indefensible commitment to preformationism, scientists avoid this vestige of a bygone age by talking about *norms of reaction* and about the issue of *heritability.* I provide an elementary explanation of these ideas in Sober (1993).
21. This is the essence of various "externalist" theories of knowledge. See, for example, Dretske (1981).
22. This, I take it, is the main conclusion of Godfrey Smith's (1991) discussion of signal detection theory.
23. Of course, the fact that different conditions pertain to *a priori* and to empirical knowledge in no way depends on the evolutionary considerations explored here. The difference is a simple consequence of applying Bayes' theorem to the reliabilist definition of knowledge.
24. Stich (1990) argues that natural selection cannot be expected to reduce the error probabilities of a perceptual device to zero. The present model reinforces Stich's point. Even if true belief is a component of fitness, it is not the whole story.
25. This is discussed by C. D. Waddington (1957) and by G. C. Williams (1966). I examine Williams' analysis of this phenomenon in Sober (1966), Section 6.2.
26. Indeed, the model developed in Moran (1992) concerning the evolution of polyphenism is essentially the same as the one presented here. However, since Moran's interests are not in cognitive strategies *per se,* the interpretive points she makes do not overlap much with the ones I have developed.
27. I use the term "measurement device" in roughly the way a physicist would. Any state of the polar bear's body that is *correlated* with the ambient temperature is a measurement device in the sense required.

REFERENCES

Courant, R., and Robbins, H. (1969). *What Is Mathematics?* Oxford University Press.

Davidson, D. (1984). "The Method of Truth in Metaphysics." In *Inquiries into Truth and Interpretation.* Clarendon Press.

Dretske, F. (1981). *Knowledge and the Flow of Information.* MIT Press.

Field, H. (1990). "'Narrow' Aspects of Intentionality and the Information-Theoretic Approach to Content." In E. Villanueva, ed., *Information, Semantics, and Epistemology.* Blackwell.

Fodor, J. (1975). *The Language of Thought.* Thomas Crowell.

Fodor, J. (1981). "The Present Status of the Innateness Controversy." In *Representations.* MIT Press.

Godfrey Smith, P. (1991). "Signal, Decision, Action." *Journal of Philosophy 88:* 709–22.

Gould, J. (1974). "Risk, Stochastic Preference, and the Value of Information." *Journal of Economic Theory 8:* 64–84.

Gould, J., and Marler, P. (1991). "Learning by Instinct." In D. Mock (ed.), *Behavior and Evolution of Birds.* Freeman.

James, W. (1897). "The Will to Believe." In *The Will to Believe and Other Essays in Popular Philosophy.* Longmans Green & Co.

Moran, N. (1992). "The Evolutionary Maintenance of Alternative Phenotypes." *American Naturalist 139:* 971–89.

Pulliam, H., and Dunford, C. (1990). *Programmed to Learn.* Columbia University Press.

Ruse, M. (1979). *Sociobiology: Sense or Nonsense.* D. Reidel.

Sober, E. (1984). *The Nature of Selection.* MIT Press.

Sober, E. (1993). *Philosophy of Biology.* Westview Press.

Stephens, D. (1989). "Variance and the Value of Information." *American Naturalist 134:* 128–40.

Stich, S. (1990). *The Fragmentation of Reason.* MIT Press.

Waddington, C. (1957). *The Strategies of the Genes.* Allen and Unwin.

Williams, B. (1972). *Morality: A Brief Introduction to Ethics.* Harper.

Williams, C. (1966). *Adaptation and Natural Selection.* Princeton University Press.

4

The primacy of truth-telling and the evolution of lying

1. INTRODUCTION

In addition to recognizing a *moral* contrast between lying and telling the truth, philosophers have sometimes postulated a *modal asymmetry.* Perhaps the most famous example occurs in Kant's ethical theory, where the moral contrast is said to derive from a modal difference. For Kant, truthfulness is right and lying is wrong *because* truthfulness can be universalized, whereas lying cannot. In his well-known discussion of promise-keeping in the *Groundwork of the Metaphysics of Morals,* Kant describes a person who needs money and is deciding whether to borrow. The question is whether it would be permissible to promise to pay back the money if you have no intention of doing so. Kant argues that morality requires that you not lie – you can make the promise only if you intend to keep it:

> For the universality of a law which says that anyone who believes himself to be in need could promise what he pleased with the intention of not fulfilling it would make the promise itself and the end to be accomplished by it impossible; no one would believe what was promised to him but would only laugh at any such assertion as vain pretense.

Kant is saying that promise-keeping could not exist as an institution, if everyone who made promises did so with the intention of breaking them. Kant's analysis of promising extends to lying in general: A world in which everyone tells the truth is possible, whereas one in which every-

I thank Martin Barrett, Ellery Eells, Malcolm Forster, Dan Hausman, Greg Mougin, Steven Orzack, Larry Samuelson, Larry Shapiro, LaVerne Shelton, Alan Sidelle, Brian Skyrms, Dennis Stampe, David Sloan Wilson, and George C. Williams for useful comments on earlier drafts. I am especially grateful to Branden Fitelson for helping me understand the issues connected with the stability of the equilibrium in the model presented here. I also benefited from discussing this paper at University of Wisconsin-Madison, MIT, and Queen's University.

one lies is unthinkable – not in the sense that it would be *bad,* but in the sense that it cannot exist.

More recently, and quite outside the discipline of ethics, Donald Davidson (1984a,b) has argued that interpreting a person's utterances is necessarily regulated by a *principle of charity.* This principle says that a correct interpretation of the corpus of utterances must have the consequence that most of what the person says is true. Davidson thinks a parallel principle attaches to *belief.* It is impossible, according to Davidson, for someone's utterances and beliefs to be mostly false.

A similar thesis has been advanced by David Lewis, who proposes that

> a language *L* is used by a population *P* if and only if there prevails in *P* a convention of truthfulness and trust in *L,* sustained by an interest in communication. (Lewis 1983, p. 169)

For a population to use a particular language is for its members to try to be truthful in what they say in that language, and for them to incline to believe the utterances of others.

Some differences among Kant's, Davidson's, and Lewis' modal claims are plain. Kant and Lewis describe a society, whereas Davidson assesses the utterances of a single individual. Kant says that lying cannot be *universal* in a society, Davidson says that falsehood cannot be *common* in the corpus of utterances that a single person produces, and Lewis says that speakers cannot usually try to assert falsehoods in the language they speak. Still, the thread that unites these three positions is that there is an asymmetry between lying and telling the truth. Glossing over some nuances, the idea is that the existence of lies entails the existence of true utterances, but not conversely. A world in which everyone tells the truth is possible, but one in which everyone lies is not.

Kant, Davidson, and Lewis view their respective theses as both *a priori* and necessary. In this paper I want to explore from a rather different point of view the idea that lying and truth-telling are asymmetrically related. I will investigate the consequences of the hypothesis that lying and credulity are behaviors that evolved by natural selection. Since this is an empirical assumption, the consequences concerning the prevalence of lying and true belief will be empirical as well. These consequences, we will see, assert that certain population configurations are *unstable,* not that they are *impossible.* If any such configuration comes to pass, selection will move the population to a new one. The claim is not that universal lying and universal false belief cannot arise, but that they will not persist.

The evolutionary ideas I'll describe are simple ones, drawn from the study of mimicry and camouflage.[1] In biology, they are usually applied to organisms that either do not have minds, or whose minds are simpler than (or at least very different from) our own. This raises the question of whether these evolutionary matters have much to tell us about lying and truth-telling in our own species. How is the protective coloration of a butterfly pertinent to a shopkeeper's dealing fairly with his customers?

It is important to recognize that an evolutionary model can be useful even when one thinks that it does not provide the correct explanation of the phenomenon at hand. The model has the virtue of making explicit a set of assumptions that jointly suffice to predict the behavior in question. If one rejects the explanation, the challenge is to say which of the assumptions is mistaken. But that should not be the end of the matter. Since all models contain oversimplifications, it will not be hard to find assumptions that are false. The real challenge is to see whether new models can be constructed that make quite different predictions.[2]

2. VERNACULAR LYING AND EVOLUTIONARY LYING

In common parlance, people lie when they say something false with the intention to deceive. Innocent error isn't lying. And accidentally saying something true when one wanted to deceive isn't lying either.[3]

What would an evolutionary concept of lying look like? Since the vernacular concept entails *intentional* deception, I suggest that the evolutionary concept should require deception *by design*.[4] The mimicry of the Viceroy butterfly is a form of evolutionary lying. The Viceroy is lying when it looks like a Monarch, because its pattern of coloration evolved for a certain reason. The coloration evolved because it causes birds that might eat Viceroys to think that the Viceroy has a nasty taste when in fact it does not.

Evolutionary lying is an historical concept; it describes why the trait evolved and does not say anything about the trait's current effects.[5] Suppose the howling of wolves evolved because howling is a way for wolves to communicate their locations to each other. However, when wolves howl in Transylvania, this causes human beings to think that Dracula is nigh. This reaction by human beings benefits the wolves, since human beings are thereby less likely to kill wolves or to interfere with their hunting. Still, the howling is not an example of evolutionary lying. The howling *causes* a false belief, but it did not evolve *because* it has that consequence.

According to the present proposal, a perfectly mindless organism can be an evolutionary liar. However, lying evolves only when there are *consumers* of lies who may be misled. Evolutionary liars must interact with organisms who form beliefs (or something like beliefs), even if the liars themselves form no beliefs at all.

If evolutionary lying is given the historical interpretation I've suggested, then symmetrical requirements should be placed on the concept of evolutionary truth-telling. A sentinel crow is telling the truth when its warning cry alerts conspecifics to the presence of a predator. But a fox is not telling the truth when its pawprints in the snow inform the farmer that the hen house is in peril. To be sure, the pawprints *cause* a true belief, but foxes don't leave pawprints *because* this helps farmers (or anyone else) to ascertain their whereabouts.

3. BATESIAN MIMICRY

Monarch butterflies (*Danaus plexippus*) are poisonous to blue jays. If a blue jay eats a Monarch, the bird will vomit. What is more, blue jays are able to learn from experience; they associate the revolting experience that follows a Monarch meal with the Monarch's striking visual appearance. When presented with a new Monarch, blue jays will refuse to eat.

The standard evolutionary explanation is that the Monarch evolved its striking coloration as a warning sign. The gaudy coloration evolved because this led predators to realize that Monarchs are bad to eat. The Monarch is, in the sense defined before, an evolutionary truth-teller. Its appearance means "I am nasty to eat," which in fact is true.

Viceroy butterflies (*Limenitis archippus*) are mimics and Monarchs are their models. The Viceroy is not poisonous to blue jays, but the Viceroy and the Monarch have remarkably similar appearances.

Viceroys are evolutionary liars. Their visual appearance says "I am nasty to eat," but this isn't so. Viceroys evolved their visual appearance because this appearance leads blue jays to form certain false beliefs.[6]

4. WHEN DOES IT PAY FOR A BLUE JAY TO BELIEVE?

When a blue jay sees an organism with the characteristic appearance that Viceroys and Monarchs share, the bird faces a problem. It has to decide whether the butterfly is telling the truth when it says "I am nasty to eat."[7]

If the bird believes the signal, the jay will not eat. This is the outcome to the jay, regardless of whether the signal happens to be true or false. On the other hand, if the bird *dis*believes the signal, its payoff crucially depends on whether the signal is true. If the butterfly is indeed nasty to eat, and the bird believes that this isn't so, the bird will eat and then regurgitate. On the other hand, if the butterfly is palatable (despite its misleading coloration) and the bird realizes that this is so, the bird will obtain a nutritious snack.

The consequences for the bird if it BELIEVES or DISBELIEVES the butterfly's signal depend on whether the signal is True:

		Sender's signal ("I am nasty to eat") is	
		True	False
Receiver	BELIEVES	$x + b_1$	$x - c_2$
	DISBELIEVES	$x - c_1$	$x + b_2$

In the example under discussion, $b_1 = c_2 = 0$; as I mentioned, the bird abstains from eating if it believes the signal, so its payoff is the same whether the butterfly is palatable or not. However, for the sake of generality, we will leave open what values the two parameters should be assigned.

If p is the probability that the sender's signal is true, the fitness of believing the signal (BEL) and the fitness of disbelieving it (DIS) are

$$w_{\mathrm{BEL}} = p(x + b_1) + (1 - p)(x - c_2).$$
$$w_{\mathrm{DIS}} = p(x - c_1) + (1 - p)(x + b_2).$$

We now can calculate when believing the signal is fitter than disbelieving it:

$$w_{\mathrm{BEL}} > w_{\mathrm{DIS}} \text{ if and only if } p(b_1 + c_1) > (1 - p)(b_2 + c_2).$$

Let us adopt the assumption that true belief is better than falsehood – that is, that $(x + b_1) > (x - c_1)$ and that $(x + b_2) > (x - c_2)$.[8] This is plausible enough in the case of our blue jays, and its more general significance will be discussed in a while. In any event, this assumption allows us to rearrange the criterion for the evolution of blue jay gullibility:

$$(1) \quad w_{\mathrm{BEL}} > w_{\mathrm{DIS}} \text{ if and only if } p > (b_2 + c_2)/(b_1 + c_1 + b_2 + c_2).$$

The assumption that true belief is better than false belief ensures that the ratio of costs and benefits in proposition (1) falls between 0 and 1.

So whether believing or disbelieving is better depends on the frequency of truth-telling. If Monarchs are common and Viceroys are rare, then most of the organisms with the characteristic visual appearance will be unpalatable. This means that truth-telling predominates – p is high. This makes it easier for gullibility to evolve. On the other hand, if Viceroys are quite common, it may no longer be true that jays who believe the signal do better than jays who do not.

5. WHEN DOES IT PAY FOR A VICEROY TO LIE?

Viceroys lie by mimicking the coloration that Monarchs possess. If the blue jay is going to believe what the butterfly says, then the Viceroy is better off saying that it is poisonous; in this instance, lying is better than telling the truth. On the other hand, if the blue jay is going to *dis*believe the butterfly's message, the Viceroy should tell the truth; it should be plain rather than gaudy. So the payoffs to the Viceroy if it LIES and if it TELLS THE TRUTH are

		Receiver	
		Believes	Disbelieves[9]
Sender	LIES	$y + b_3$	$y - c_4$
	TELLS THE TRUTH	$y - c_3$	$y + b_4$

If q is the probability that the receiving blue jay will believe the message, the fitnesses of lying (LIE) and of truth-telling (TT) are

$$w_{\text{LIE}} = q(y + b_3) + (1 - q)(y - c_4)$$
$$w_{\text{TT}} = q(y - c_3) + (1 - q)(y + b_4)$$

We now can calculate when lying is fitter:

$$w_{\text{LIE}} > w_{\text{TT}} \text{ if and only if } q(b_3 + c_3) > (1 - q)(b_4 + c_4).$$

If we adopt the assumption that it is better for Viceroys to lie than to tell the truth when jays are inclined to believe that gaudy individuals are poisonous, but better for Viceroys not to lie when jays are inclined to believe that gaudy individuals are *not* poisonous [i.e., $(y + b_3) > (y - c_3)$ and $(y + b_4) > (y - c_4)$], we may rearrange this criterion:

$$\text{(2)} \quad w_{\text{LIE}} > w_{\text{TT}} \text{ if and only if } q > (b_4 + c_4)/(b_3 + c_3 + b_4 + c_4).$$

If most jays are gullible, q is large, which makes it easier to satisfy the condition needed for Viceroys to find an advantage in lying. On the other hand, if most jays disbelieve the signals they receive, it may not be worthwhile for a Viceroy to expend the energy required to construct the mimetic appearance.

6. APPLYING THE MODEL TO INTERACTIONS WITHIN A SINGLE SPECIES

The example I have been discussing concerns relations among three species – the Monarch is the model, the Viceroy is the mimic, and the blue jay is the so-called *operator.* In each of the two mathematical models, I discussed the evolution of a pair of traits *within* a single species. Blue jays evolve their inclination to believe or disbelieve the hypothesis that a particular pattern of coloration indicates disgusting flavor. And Viceroy evolution takes the form of mimetic coloration replacing some hypothetical "plain" appearance.

Even though this example involves three species, it breaks no conceptual rules to apply the same formal model to a case of evolution within a single species. As it happens, within-species mimicry exists; it is called *automimicry.* So let us imagine a single population, in which individuals are both senders and receivers of signals. Each organism exhibits a pair of traits. One trait characterizes whether the organism believes or disbelieves the messages it receives from conspecifics; the other trait indicates whether the organism is a liar or a truth-teller with respect to the messages it transmits.

A model of the joint evolution of these traits may be obtained simply by conjoining propositions (1) and (2), if we assume that an organism's probability of being gullible is independent of whether it is a liar or a truth-teller. Just for the sake of exploring a simple model, let us adopt this assumption.[10]

We need to be clear on what this model describes. We are imagining that the population begins with a *correlation* in place; an organism is tasty if and only if it is plainly colored, and it is poisonous if and only if its coloration is gaudy. In this initial configuration, the population exemplifies only two of the four possible combinations of characters. In principle, we could pose two questions about how the population will subsequently evolve:

(a) Under what circumstances will the population evolve to a configuration in which plain organisms are not always tasty?
(b) Under what circumstances will the population evolve to a configuration in which gaudy organisms are not always poisonous?

My discussion of the Viceroy/Monarch/blue jay system focused on question (b), not question (a). A full treatment of the joint evolution of lying/truth-telling and of belief/disbelief would have to address both questions. However, for the sake of keeping matters simple, I will con-

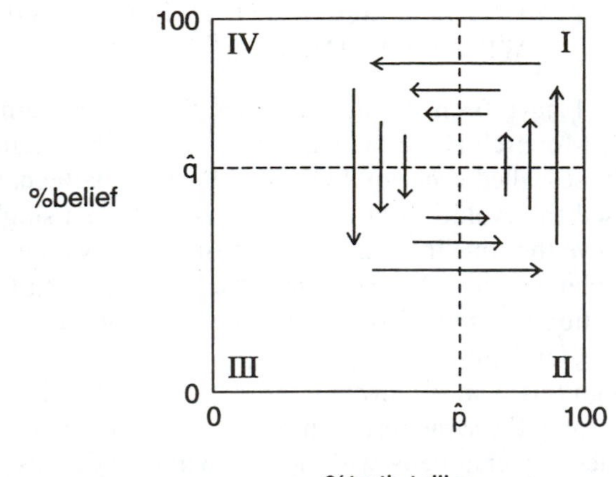

Figure 4.1

tinue to limit myself to question (b).[11] In this setup, plain individuals never lie; they are always tasty and the operator knows this. However, gaudy individuals may be lying or telling the truth, and so the operator who confronts a gaudy individual faces a problem of decision making under uncertainty.

Propositions (1) and (2) involve "coupled" criteria; the criterion for the evolution of each pair of traits depends on the frequency of the other. We can represent the population's evolution as a trajectory through a two-dimensional space, one dimension representing the frequency of belief/disbelief, the other representing the frequency of lying/truth-telling, as depicted in Figure 4.1.

We now must ask whether there are any points in this space that are stable equilibria. This means that the population, if it reaches that configuration, will not depart from it. I'll begin with some negative consequences. *No point along the edges of the unit square is stable.* If a population begins with 100 percent gullibility, it will depart from that homogeneous state. The reason is that if everyone is gullible, lying eventually will evolve, and this will provide a reason, sooner or later, for disbelief to start displacing gullibility. The same is true of 100 percent disbelief. If no one believes lies, the energetic cost of lying will not be worth paying, so truth-telling will start to increase. But once truth-telling becomes common, it will no longer pay to be mistrustful. The configurations of 100 percent lying and 100 percent truth-telling are also unstable. If everyone tells the truth, gullibility will evolve, and this

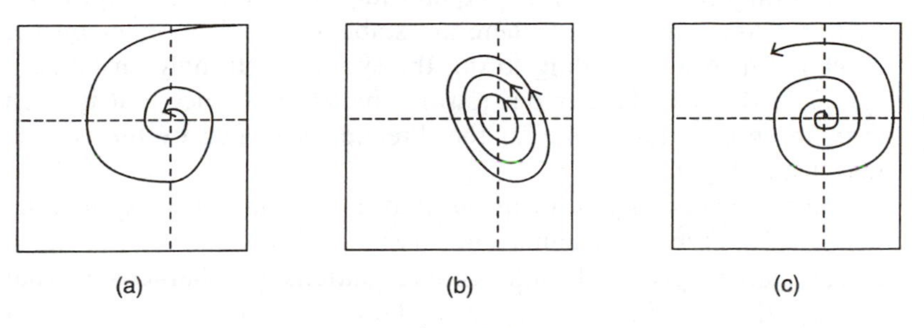

Figure 4.2

will provide an incentive for individuals to lie. Similarly, if everyone lies, mistrust will evolve, which will remove the advantage of lying.[12]

If a homogeneous population isn't stable, is there any point in the interior of the unit square that is a stable equilibrium? There *is* a unique equilibrium point, which I'll describe. But whether it is stable or unstable is a further matter.

An equilibrium is defined as a point at which lying and truth-telling are equal in fitness as are belief and disbelief. The location of this point depends on the costs and benefits associated with the traits. Proposition (1) says that belief and disbelief are equally fit with the frequency ($\hat{p}$) of truth-tellers equals $(b_2 + c_2)/(b_1 + c_1 + b_2 + c_2)$. Proposition (2) says that lying and truth-telling are equally fit precisely when the frequency ($\hat{q}$) of believers equals $(b_4 + c_4)/(b_3 + c_3 + b_4 + c_4)$. So $(\hat{p},\hat{q})$ is the unique equilibrium. Notice that if all costs and benefits are positive, this point is in the square's interior, not on any of its edges.[13] Moreover, it could lie *anywhere* in the square's interior, depending on what the costs and benefits happen to be.

In Figure 4.1, I've drawn arrows to indicate the vector components of change that the population experiences. For example, in quadrant I, when truth-telling and belief both exceed their equilibrium values, the population will move north and west. The effect of this is that truth-telling will decline in frequency. In quadrant IV, the vectors move the population west and south. And so on.

If there is just one equilibrium point in the population's evolution, we now must ask whether that point is *stable.* This means that if the population is in the neighborhood of the equilibrium, the population will move toward it. The vectors depicted in Figure 4.1 do not answer this question. In principle, they leave open which of the three possible trajectories depicted in Figure 4.2 the system will display.

According to option (a), the equilibrium point is a stable spiral attractor.[14] According to (b), there are stable orbits that center on the equilibrium. And according to (c), the system is strongly unstable; it spirals farther and farther out toward the edges. Notice that in each case, the population moves in the direction specified by the vectors depicted in Figure 4.1.

Interestingly enough, it turns out that the stability of the system depends on the type of dynamics one assumes. If change is assumed to be continuous, specified by differential equations that characterize what is called the Darwin flow, then (a) is the result. On the other hand, if change is discrete and is characterized by the so-called Darwin dynamics, (c) is the consequence. And if the continuous dynamics is given by the so-called replicator equations, then (b) is what occurs.[15] This means that the issue of stability cannot be resolved until an empirical argument is provided concerning which type of dynamics is most plausible for the evolving system one wishes to discuss.

Fortunately, these fine points don't matter as far as the purpose of this paper is concerned. For regardless of which dynamics one considers, the point of importance is that there is nothing privileged about the upper-right hand corner of Figure 4.1. In case (a) depicted in Figure 4.2, a preponderance of truth-telling and trust will be stable for some assignments of costs and benefits, but unstable for others. In case (b), the population will retain a preponderance of truth-telling and trust only for some assignments of payoffs and initial trait frequencies. And in case (c), a preponderance of truth-telling and trust is always unstable, regardless of what the payoffs are. In summary, *if a preponderance of truth-telling and trust is stable, its stability depends on some highly contingent properties of the evolutionary process.*

7. CONSEQUENCES

In the light of this conclusion, what becomes of Kant's claimed asymmetry? He said that universal lying is impossible, but that universal truth-telling is possible. At most, what we find in this model is that neither is *stable.* There is no formal reason why a population cannot (perhaps by chance) at some time be such that everyone lies or everyone tells the truth. The point is that such configurations, if attained, will not long endure.[16]

Still, there is an asymmetry between lying and truthfulness, but it involves their historical relationship, not their possible universality. *For mimicry to evolve, there first must be a model.* The Monarch evolved its warning coloration before the Viceroy evolved its mimetic appearance. The Monarch had an evolutionary reason to tell the truth, whereas the Viceroy had an evolutionary reason to lie. Could the historical sequence have occurred in the reverse order? Could the Viceroy have evolved its deception before the Monarch evolved its truthfulness? The answer is *no,* because blue jays then would have had no reason to interpret the appearance of the Viceroy as indicating something that isn't true. In the absence of the Monarch's truth-telling, blue jays would have come to associate the Viceroy's appearance (whatever it happened to be) with the presence of a palatable meal. In this circumstance, the Viceroy's appearance would have meant "I am good to eat," which is no lie.[17]

If truth-telling is given the historical interpretation suggested in Section 2, it isn't quite correct to say that the evolution of lying requires the prior existence of truth-telling. This is correct in the Viceroy/Monarch example, but it is not true in general. For consider, not mimicry, but *camouflage.* Suppose a bird (the Eurasian bittern, for example) evolves a visual appearance that makes it blend in with the reeds in which it nests. The reeds did not evolve their appearance in order to tell the truth to anyone. So evolutionary lying does not entail the existence of evolutionary truth-telling. A better formulation would be that *the evolution of lying requires the prior existence of (at least some instances of) true belief.* This seems to be a more adequate generalization about the evolution of mimicry and camouflage.[18]

Davidson says that at least half (or "most") of an individual's beliefs must be true and that at least half (or "most") of an individual's utterances must be true as well. This entails (if each agent has the same number of beliefs and if each makes the same number of utterances) that at least half of the beliefs and at least half of the utterances found in a population must be true. These consequences do not follow unconditionally from the model we are considering. It is not hard to see why. To return to our three-species example, suppose that blue jays have many food sources besides the butterflies we have been considering. If so, they may be better off passing over a gaudy butterfly rather than running the risk of getting sick from eating it. This fact about the jays can lead mimetic Viceroys to become more numerous than the Monarch models. The result is that even though most of the individuals with the visual appearance shared by Viceroys and Monarchs are lying, the jays have no evolutionary reason to become less trusting. In this case, when a blue jay believes "this butterfly is nasty to eat," the bird will usually be mistaken. Both truth-telling and true belief will occur less than half the time.

Of course, in the real world, Viceroy mimics are less common than the Monarchs they imitate, and this pattern often obtains in systems of Batesian mimicry. However, the reason for this is entirely contingent on the costs and benefits experienced by the organisms in question. Blue jays will be choosy about what they eat when they have lots of choices. But when the pickings are slim, a possible parcel of nutrition will be more worth the risk of getting sick.

In assessing Kant's and Davidson's modal asymmetries, it is essential to separate *synchronic* from *diachronic* questions concerning the primacy of truth-telling. *At a given time,* there is no reason why most of the utterances or beliefs found in a population must be true. In fact, there is no theoretical stricture against *all* beliefs and *all* utterances being false, *at a given time.*[19] However, should such configurations presently obtain, the model under discussion tells us something about the past and the future. In the prelapsarian beginning, there was at least some true belief. And if universal lying beclouds the present, the future probably will be brighter.

So the model does not guarantee that lying is less than universal *now.* Still, biologists who employ models like the present one usually expect natural populations to be at or near equilibrium. Biologists who think this way are making the empirical and contingent assumption that the postulates of the model have been true for a very long time – for long enough for the population to have evolved to its predicted equilibrium.[20]

In summary, there is reason to think that universal lying and universal false belief were not the *initial* state of the population. However, it is not impossible that *present* populations should exhibit this peculiar configuration. When biologists assume that this isn't so in a contemporary population, they are viewing universal lying as an empirical improbability, not as an *a priori* impossibility.

As noted earlier, Davidson's theses about belief and utterance are stated in terms of the mix found in a single individual. However, the model I have described involves just one *type* of utterance and one *type* of belief. To be sure, a single organism can form many *tokens* of that one type. For example, each time a jay confronts a gaudy butterfly, the bird must decide whether the butterfly is palatable. So, over the jay's lifetime, it will form numerous token beliefs of a given type. But still, the bluejay/butterfly system involves a quite limited repertoire of sender and receiver states. How might the model be applied to organisms who do more?

The model can be applied straightforwardly to this more complex situation. For each proposition about which the receiver must form an opinion, criterion (1) may be applied. Of course, the costs and benefits may differ from problem to problem, but the basic form of the model remains the same. Senders who dispatch a multiplicity of messages may similarly be modeled by repeated applications of criterion (2). Depending on the details of the problems involved, the consequence is likely to be that a receiver should be trusting about some propositions, but quite distrustful about others. And a sender should sometimes lie and sometimes tell the truth. In this model, each organism has many beliefs and makes many utterances. But, as before, there is no unconditional guarantee that most of an individual's beliefs and utterances will be true.

A further complication that is worth considering is that organisms often do not have a separate mechanism to cope with each proposition about which a decision must be made. Here I am alluding to the *systematic* quality of cognition and of problem-solving devices generally.[21] This means that different problems of interpretation will not be independent of each other; the solution that evolves for one problem constrains the solutions that can evolve for others. It follows that a given problem-solving policy will evolve according to its *average* performance across all the problems it addresses. This may have the result that an evolved strategy may do quite poorly with respect to some of the problems it confronts.

My point in mentioning the complication of multiple problems and the fact of systematicity is that neither of these details can be expected to rescue Davidson's thesis. The basic model comprised of propositions (1) and (2) does not ensure that more than half (or "most") of the beliefs and utterances that occur in a population will be true. This result will not be reversed by making the model more realistic in the two ways just mentioned.

Similar remarks apply to Lewis' view, although his use of the concept of convention raises some special issues. For Lewis, conventions are sustained by the "common interests" of the participants; the concept applies to cases "in which there is a predominant coincidence of interest," rather than to "cases of deadlocked conflict" (Lewis 1983, p. 165). This remark helps explain why the model explored here conflicts with the framework that Lewis adopts. Mine is a model of *individual* selection, in which the different organisms in a single population are in competition with each other.[22] It is hardly to be expected in that setting that individuals will always evolve the inclination to tell the truth or to believe what others say. Of course, if we shift to a group selection framework, matters may change. If speech communities compete with each other, it may sometimes be true that populations whose members are trusting and truthful do better than populations whose members exhibit distrust and deceit. But, once again, the predominance of trust and truthfulness turns out to be contingent and empirical, not necessary and *a priori.*

In fairness to both Davidson and Lewis, I should explain more carefully how they would take issue with the position I've been developing. Lewis considers the objection to his view that there could be "a population of inveterate liars, who are untruthful in L more often than not" (Lewis 1983, p. 182). His reply is to deny that L is in fact the language that this community is using. Rather, the community is using a different language, L', which contains the same sentences but assigns them different truth conditions. Individuals are speaking L', and they are conforming to the convention of trust and truthfulness in L'. Davidson, of course, has the same reaction, except that he focuses on a single individual rather than on a community, and his claim concerns the *actual* truth of the utterances, not the speaker's *trying* to say what is true. For Davidson, if your assignment of meanings to a speaker's utterances entails that the speaker is mostly mistaken, then you must modify your interpretation of what the utterances mean.

Let us apply Lewis' interpretive strategy to the mimicry example of this paper. Imagine a case in which Viceroys outnumber Monarchs, but blue jays still abstain from eating butterflies with the distinctive coloration that the butterflies share. There are parameter values of the model and dynamical assumptions that entail that this arrangement will evolve. We would seem to have here a case in which most of the butterflies are "inveterate liars." Lewis, however, would have us reinterpret what the butterflies' distinctive coloration means. It cannot mean "I am nasty to eat," because that is usually false. Rather, it must mean something true. But what?

One possibility is that the gaudy coloration means "I am good to eat." However, if this is what the coloration means to the jays, and if jays are trusting, as Lewis suggests, then it is very strange that jays avoid eating such butterflies. By interpreting the *senders* of messages as mainly truthful, we are saddled with a puzzle concerning why *receivers* of messages behave as they do.

A further difficulty with this suggested interpretation arises when we consider what blue jays think as mimics increase in frequency. Suppose, when models vastly outnumber mimics, that blue jays think that gaudy butterflies are poisonous. But now suppose that the mimics start increasing in frequency relative to the models. Jays, we are imagining, do not change their behavior. It is an entirely natural hypothesis that they continue to think what they thought before. In this process, no magical change need occur in the jays when mimics start outnumbering models. There is no reason to hold that jays must have one belief about gaudy butterflies when the mimics are at 49 percent and a quite different belief when mimics are at 51 percent.

I've just considered one way to attribute mostly true beliefs to jays. But, of course, other interpretations are conceivable. Recall that the case we are considering is one in which mimics outnumber models, but jays abstain from eating either. Let us take up a new suggestion – that jays interpret the coloration to mean "The probability that this butterfly is poisonous exceeds 0.10." On this reading, the jay believes something true about the gaudy butterfly it beholds, and the jay abstains from eating because it would rather be safe than sorry. What is more, this proposition remains true as mimics evolve from being rare to being quite common.

A weakness in this hypothesis becomes visible when we consider a *transplant experiment*. Let us whisk this jay into another valley, one in which gaudy butterflies are poisonous *less* than 10 percent of the time. Surely it is reasonable to hold that the jay's belief does not instantaneously change. The jay will interpret the first gaudy butterfly it sees in the new valley in the same way it interpreted the last gaudy fly it saw in the old locale. After all, the jay abstains from eating in its new home, just as it did before, and the relocation process gave it no new information. If the jay sees a few more butterflies in the new valley than it saw in the old one, it will usually be mistaken if it believes, each time, that the butterfly it beholds has a probability greater than 0.10 of being poisonous.[23]

Other interpretations that attribute mostly true beliefs to blue jays could be considered as well, but my hunch is that they will be no more successful. In addition, it is well to remember that Lewis' thesis demands not just that such interpretations can be invented, but that they are *superior* to the hypothesis I have suggested. My feeling is that biologists are on firm ground when they interpret mimicry as an instance of deception. This intuitive biological idea does not crumble into incoherence when mimics happen to be more common than models.

8. CONCLUDING COMMENTS

My discussion of the Viceroy/Monarch/blue jay example may suggest that my argument is committed to a specific theory of meaning, which, once stated, will be vulnerable to objections. However, I doubt that the argument requires anything terribly substantive by way of a semantic theory. Of course, I *am* committed to the idea that mimicry involves *deception.* But beyond this, I don't care what specific content one assigns to the Viceroy's appearance. I have construed it to mean "I am nasty to eat," but the argument would have worked just as well if the content had been "If you eat me, you will throw up."

Even if I have not assumed any full-blown theory of content, it may seem that I at least have bought into a substantive necessary condition. Whether or not the Viceroy's appearance means "I am nasty to eat" or "If you eat me, you will throw up," it seems that the butterfly's appearance has this content because of why Viceroys evolved their distinctive coloration. If so, content is constrained by etiology.

Although this proposal about meaning may be defensible, I do not think I need it for my present argument. Perhaps the Viceroy's appearance has the meaning it does because of its history. But, then again, maybe the content derives from facts concerning what blue jays now do when they see the pattern of coloration. Again, I don't care *why* the Viceroy's appearance means what it does, only that it *has* a meaning such that blue jays end up with false beliefs when they interpret the coloration.

Philosophers who accept the views of Davidson or Lewis may be inclined to see a yawning chasm between my birds and butterflies on the one hand and the rich fabric of human belief and assertion on the other. It would be difficult to dispel completely this suspicion without providing an adequate account of belief and meaning, something I certainly am in no position to do. However, in the absence of such a defense, let me suggest a *diagnosis* for why Davidson's principles of charity and related notions have the *prima facie* plausibility they do.

Suppose a naive naturalist, untutored by Darwinian theories of deception, happens on a firefly in a forest. The organism produces a flashing light, a behavior that seems to play some communicative function. If the firefly's flashing communicates *something,* what does it mean? A reasonable first guess might be that the meaning of the flashing is some state with which the flashing is correlated. Just as smoke means fire, so the light means something that is usually true when such flashes occur and false when they do not. So the naturalist guesses that the flash means "I am here" and functions as a mating signal for conspecifics. If this is what the signal means, it is true each time the signal is sent. This is a *simple* hypothesis. Perhaps it is the default hypothesis that should be favored in the absence of specific evidence to the contrary.

A radical interpreter, who aims to interpret expressions in an alien tongue, proceeds in similar fashion. If informants seem deliberately to produce the noise "gavagai" on certain occasions, it may be plausible to think that this noise is an *utterance.* But if "gavagai" communicates something, what does it mean? If informants usually say "gavagai" when rabbits are present and do not say this when rabbits are absent, it may be a good first guess that "gavagai" means *rabbit* (or something coextensive). The hypothesis that utterances are usually true may be a reasonable default hypothesis that we should adopt unless we have evidence to the contrary.

For both the naturalist and the radical interpreter, what is initially credible may later be discarded when new evidence is acquired. The naturalist may learn that this species of firefly attracts fireflies of *other* species, which it promptly eats. It manages to do this because its flashing is a passable copy of the mating signal that the other species produces. The conclusion is then drawn that the firefly under study is a mimic; each time the predator firefly emits this signal, it is lying.[24]

The initial hypothesis formulated by the radical interpreter can suffer a similar reversal of fortunes, once further evidence is adduced. Suppose the linguist investigates the religion of the culture under study, and finds that rabbits are thought to be a sacred food provided by the goddess of rabbits. This and other evidence may lead to the conclusion that "gavagai" is not the name of anything earthly, but refers to the goddess herself. For both the naive naturalist and the radical interpreter, the initial, default hypothesis is a ladder that may be kicked away once it has been climbed.

It is ironic that Davidson has grounded his defense of the principle of charity on the idea that interpretation is *holistic.* For, on at least one interpretation of that slippery term,[25] holistic interpretation means that a complex set of data must be used to judge a number of interlinked hypotheses simultaneously. It is epistemically quite significant that we do not solve the problem of interpreting the firefly's flashing or the informant's utterance of "gavagai" in isolation from other information, once that further information becomes available. It is this interanimation of further considerations that can undermine one's initial inclination to interpret all or even most utterances as true. Holism, so interpreted, tells *against* the principle of charity.

A last objection to my argument is worth considering. In the example of Batesian mimicry, the sending and interpreting of signals is under the control of natural selection. But human communication is a product of culture; the shopkeeper does not lie to his customers because natural selection provided him with a gene that makes him do so.

Although I concede that many features of human communication are a product of culture,[26] I still believe that the model discussed here has some bite. For let us interpret the model as describing cultural evolution, not biological evolution.[27] In many cultural contexts, lying and truth-telling, as well as trust and distrust, are designed to minimize costs and maximize benefits. There is no reason why these costs and benefits must be calibrated in the currency of survival and reproductive success. For example, an economic market is populated by buyers and sellers. The sellers make claims about their products; the buyers must evaluate whether the claims are true. It is not just in biological evolution that lying is most apt to succeed when it is rare, nor is it solely in the wild that credulity is regulated by the maxim of *caveat emptor.* Even if cultural change has become autonomous from the process of biological evolution, a model of Batesian mimicry may have something to tell us about the primacy of truth-telling.

Returning, in conclusion, to the gap that allegedly separates human beings from the rest of nature, we should ponder the implications of this chasm. Davidson denies that nonlinguistic creatures have beliefs, and his definition of a language is stringent enough to ensure that human beings are unique in this respect. If this is right, then my talk of mimicry in terms of deception – of receivers interpreting messages and thereby forming false beliefs – must be viewed as so much metaphor. *So be it.* Let us, for the sake of argument, describe the mimicry system by the terms assertion* and belief*, so as not to confuse the real thing with its alleged counterfeits.[28] Even so, it remains true that properties of the human system of belief and assertion are descended from properties of the system of belief* and assertion* used by our ancestors. Such "primitive" signaling systems are the precursors of what we find among human beings today. With respect to this ancestral signaling system, we have seen that the assertions* and beliefs* found in a population can be mostly false. Davidson must argue that somewhere in the evolution of human belief and assertion, matters changed. This is not inconceivable, of course. But its mere conceivability is no reason to think that it is plausible. Those persuaded that the principle of charity is correct must do more than insist on its status as an *a priori* truth. They must show why the transition from belief* and assertion*, for which the principle fails, to belief and assertion, for which the principle allegedly holds, should have occurred. The theoretical challenge is to show why the principle should emerge in the natural history of systems of communication. This, I believe, will be no small task.

NOTES

1. The rather straightforward biological details I'll describe are discussed, for example, in Owen (1980).
2. That is, the real issue is whether various consequences of the model are *robust,* not whether all the assumptions behind the model are *true.* Levins (1966) emphasizes the importance of finding "robust theorems"; see Orzack and Sober (1994) for discussion.
3. This second clause in the definition of lying may be less obviously correct than the first. Including it won't affect the main conclusions I draw, however.
4. My talk of "design" here does not entail the existence of an intelligent designer. When evolutionists speak of organisms "being designed" to have a trait, they mean that the feature evolved because there was a selective advantage in having it.
5. Here I follow Williams' (1966) idea that *adaptation* is an historical concept. See Sober (1993) for further discussion.
6. The Monarch/Viceroy/blue jay system is an example of so-called Batesian mimicry, named after the nineteenth-century naturalist H. W. Bates. In Batesian mimicry there is a palatable mimic and an unpalatable model. In so-called Mullerian mimicry, both mimic and model are unpalatable.
7. For the sake of convenience, I take the jay to have two options. This means that not believing a proposition is equated with disbelieving it. A model in which there are three possible responses (belief, disbelief, and suspending judgment) is possible, as well as one with more options still. Some aspects of the three-option case were explored in Essay 3.
8. The quantity $(b_1 + c_1)$ represents the *importance* of the proposition "I am nasty to eat." It indicates how much it matters whether the receiver believes or disbelieves that proposition, if it is true. See Essay 3 for further discussion of this concept. The assumption that true belief is better than false belief is the idea that the propositions in question have *positive* importance.
9. Suppose the jay's options are believing the message or *ignoring* it. If the jay ignores the message, the Viceroy is better off avoiding the energetic costs of constructing a gaudy coloration. The inequalities among the pay-

offs in this case are the same as the ones that obtain when the jay either believes or disbelieves.

10. Within our own species, do liars tend to be less gullible than truth-tellers? The answer isn't obvious. Although truth-tellers are sometimes naive, liars are sometimes self-deluded.
11. Greg Mougin has worked out the details of the more general model, in which both signals can be misleading. The results described for the simpler model obtain in this more general setting.
12. These arguments presuppose, as is usual in the analysis of equilibria in evolutionary models, that an input of mutations is available to "test" the stability of homogeneous configurations.
13. The only way to avoid this consequence within the present model is to withdraw the assumption I mentioned in deriving propositions (1) and (2) – that the costs and benefits are such that, for each i, $(b_i + c_i) > 0$. For example, if blue jays are better off believing that a gaudy butterfly is nasty to eat, *regardless of whether it is so in fact,* then universal credulity can evolve. In this case, $(x + b_2) < (x - c_2)$. Likewise, if it is better for a butterfly to be plain than gaudy, *regardless of how receivers interpret their appearance,* then universal truth-telling can evolve. In this case, $(y + b_3) < (y - c_3)$. But barring such eventualities, the population equilibrium must be a mixed state.
14. When a stable equilibrium involves population heterogeneity, the question remains of how the equilibrium will be realized. For example, it might mean that n percent of the individuals are liars and $(100 - n)$ percent are truth-tellers, or it may mean that each individual lies n percent of the time and tells the truth $(100 - n)$ percent of the time. For discussion of the difference between these two formats, see Orzack and Sober (1994).
15. These three dynamics are specified by equations (27.1), (27.2), and (27.3) of Hofbauer and Sigmund (1988, p. 273). See pp. 274–283 for the stability results. Skyrms (1990, pp. 64, 176) has done simulations that support the claim that the continuous dynamics applied to the family of games described here entails that the equilibrium is a stable spiral attractor.
16. Michael Ruse has anticipated the present conclusion. In discussing Kant's attempt to show that the "Imperative's violation leads to 'contradictions,'" Ruse observes that "these are hardly literal logical contradictions, but are instead the kinds of maladaptive interactions that the Darwinian would think eliminated by selection" (Ruse 1986, p. 262).

    In fairness to Kant, it should be noted that the passage from the *Groundwork* quoted at the beginning of this essay seems to address the question of whether universal lying is stable, not whether it is synchronically possible. Of course, if Kant's point is that universal lying is unstable, the claimed asymmetry dissolves, since the same is true of universal truth-telling.

    Kant frequently formulates the universalizability criterion disjunctively;

he talks about whether an action could be universalized and also about whether a "rational agent" could will that the action be universal. Even if my argument shows that universal lying is possible, the question remains of whether a rational agent could will that everyone always lie.

17. This argument about the primacy of truth-telling within the context of the evolution of mimicry assumes that the operator (the jay, in this case) is better off with true beliefs than false ones. For discussion of this assumption, see Essay 3.
18. Of course, nothing so cognitive as belief on the part of the operator is needed for evolutionary lying to evolve. Pared closer to its minimum, the requirement is that the operator first have evolved the behavioral regularity to produce action *A* in circumstance *C* because doing *A* in *C generally* produces benefit *B*. The mimic then is able to cause the operator to do *A* in *C* even though doing *A* in *C* does not in *those* cases produce *B* for the operator (though it does produce some other benefit for the mimic). In this format, we may say that *C* "indicates" to the operator that *A* will yield *B*, but here indication clearly carries no requirement of mental representation.
19. The synchronic possibility does not mean *at a split second instant*. The moment of time can be "thick." What is *possible* for a second also is *possible* (though very *improbable*) for a thousand years.
20. See Maynard Smith (1978) and Sober (1993) for discussions of this and other assumptions that underlie the application of optimality models.
21. I discussed this point in connection with the evolution of learning and *a priori* prejudice in Essay 3.
22. See Sober (1993) for an introduction to the idea of individual selection and the problem of the units of selection in which it figures. These issues also are discussed in Essays 1 and 7 of the present volume.
23. The idea of a transplant experiment offers a general format for constructing an objection to the Davidsonian position. Let the Davidsonian stipulate what an agent's beliefs are *in a given environment*. Then whisk the agent to a new environment in which the agent's stipulated beliefs would mainly be false if they are the same as they were in the old environment. The Davidsonian is thus committed to the agent's beliefs having changed – a commitment, I believe, that will be hard to justify.
24. The example is not invented. Females of the genus *Photuris* are the predatory mimics I am describing; they prey on males of the genus *Photinus*. It also is worth noting that the firefly example involves a mimetic *behavior*, whereas the Viceroy example involves a mimetic *morphology*. Even if one insists that lying must be a behavior, the theory under discussion has plenty of applications to mimetic behaviors.
25. For discussion of its multiple meanings, see Fodor and Lapore (1992).
26. Of course, *all* human communication requires the use of a genetically specified set of characteristics. The point is that many *differences* in com-

municative behavior are not to be explained as due to genetic differences, or as fitness maximizing. See Sober (1993) for further discussion.

27. See Sober (1993) for discussion of the relation between biological and cultural evolution.
28. In Essay 2, I argued that essentialist theses do not solve or dissolve questions about functional utility. The present point illustrates this more general idea.

REFERENCES

Davidson, D. (1984a). "Belief and the Basis of Meaning." In *Inquiries into Truth and Interpretation.* Oxford University Press.

Davidson, D. (1984b). "Radical Interpretation." In *Inquiries into Truth and Interpretation.* Oxford University Press.

Fodor, J., and Lapore, E. (1992). *Holism: A Buyer's Guide.* Basil Blackwell.

Hofbauer, J., and Sigmund, K. (1988). *The Theory of Evolution and Dynamical Systems.* Cambridge University Press.

Levins, R. (1966). "The Strategy of Model Building in Population Biology." *American Scientist 54:* 421–31.

Lewis, D. (1983). "Languages and Language." In *Philosophical Papers,* volume 1, pp. 163–188. Oxford University Press.

Maynard Smith, J. (1978). "Optimization Theory in Evolution." *Annual Review of Ecology and Systematics 9:* 31–56. Reprinted in E. Sober (ed.), *Conceptual Issues in Evolutionary Biology,* MIT Press, 2nd edition, 1993.

Orzack, S., and Sober, E. 1993. "A Critical Examination of Richard Levins' 'The Strategy of Model Building in Population Biology.'" *Quarterly Review of Biology 68:* 533–46.

Orzack, S., and Sober, E. 1994. "Optimality Models and the Test of Adaptationism." *American Naturalist 143:* 361–80.

Owen, D. (1980). *Camouflage and Mimicry.* University of Chicago Press.

Ruse, M. (1986). *Taking Darwin Seriously.* Basil Blackwell.

Skyrms, B. (1990). *The Dynamics of Rational Deliberation.* Harvard University Press.

Sober, E. (1993). *Philosophy of Biology.* Westview Press.

Williams, G. (1966). *Adaptation and Natural Selection.* Princeton University Press.

5

Prospects for an evolutionary ethics

1. TWO KINDS OF QUESTION

Human beings are a product of evolution. This means that the *existence* of our species is explained by the process of descent with modification. But, in addition, it also means that various *phenotypic characteristics* of our species – features of morphology, physiology, and behavior – have evolutionary explanations.

Some see in this simple idea the promise of a new understanding of the problems of ethics. Others grant its truth, but reject its relevance. The first group insists that if evolutionary biology can provide insights into the workings of the human mind, these insights cannot fail to transform our understanding of right and wrong. But for the doubters, the fact that human beings have an evolutionary past has no more significance than the fact that the human body must obey the laws of physics. It is true that our bodies are subject to the law of gravity, but apparently that is of little help in understanding why we think and act as we do.

To assess the relevance of evolutionary ideas to the problems of ethics, we must distinguish two quite different projects. The first is the task of accounting for why people have the ethical thoughts and feelings they do. The second concerns the problem of deciding what the status of those thoughts and feelings is. The difference between these two projects is illustrated by the following pair of questions:

(1) Why do people have the views they do concerning when it is morally permissible to kill?
(2) When is killing morally permissible?

I am grateful to James Anderson and Louis Pojman for useful discussion.

Problem (1) poses a problem of *explanation,* while (2) engages the task of *justification.* One of the main issues I want to consider is how these questions are related to each other. This is a matter of some intricacy. But at the simplest level, it is important to recognize that there is no automatic connection between the two types of problem.

It is quite obvious that a person can believe the right thing for the wrong reason. Consider my friend Alf, who thinks the earth is round, but does so because he thinks that the earth is perfect and that roundness is the perfect shape. Alf also happens to know various facts that the rest of us recognize as evidence for the earth's being round. But Alf's predicament is that he does not see these facts as having any particular evidential significance. Alf believes *E* and also believes *R,* but he does not believe *R because* he believes *E. E* is evidence for *R; E* is *a* reason to believe *R,* but it is not *Alf*'s reason for believing *R.*

This example encapsulates two lessons. First, it is possible to show whether a proposition is justified without saying anything, one way or the other, by way of explaining why someone happens to believe that proposition. *E* justifies *R,* but that says nothing about why Alf believes *R.* Second, and conversely, an explanation of why someone believes a proposition may leave open whether the proposition is in fact well supported by evidence. Alf's strange ideas about perfection explain why he believes *R.* But this quirk of Alf's psychology does not tell us whether *R,* in fact, is a proposition that is well supported by evidence.

So the following two questions are quite different from each other:

(3) Why does Alf believe that the earth is round?
(4) Is the proposition that the earth is round strongly supported by evidence?

I hope the parallel between problems (1) and (2) on the one hand and problems (3) and (4) on the other is suggestive. In each pair, it is possible to answer one question without thereby answering the other. In summary, we have the following slogan: *An explanation for why someone believes a proposition may fail to show whether the proposition is justified, and a justification of a proposition may fail to explain why someone believes the proposition.* This simple conclusion is the beginning, not the end, of our discussion of the relationship between such questions as (1) and (2).

2. PATTERNS OF EVOLUTIONARY EXPLANATION

Before we can address the question of how the problem of explanation and the problem of justification are related, it is well to get clear on how evolution can contribute to the former task. How might evolutionary theory help explain why we have the ethical thoughts and feelings that we possess?

Although natural selection is just one of the processes that can produce evolutionary change, it is the one of greatest interest to sociobiologists. When natural selection is cited by way of explaining why a trait is currently found in some population, an ancestral population is postulated in which the trait was one of several variants that were represented. These traits are claimed to have differed in their *fitness* – their capacity to help the individuals possessing them to survive and reproduce. Through a process of differential survival and reproduction, the population moved from a mixed condition to the homogeneous condition we now observe. Why do zebras now run fast when a predator approaches? Because, ancestrally, zebras differed in running speed, and the fast zebras survived and reproduced more successfully than the slow ones.

Within this basic outline, it is important to distinguish two quite different patterns of explanation that may be used to elaborate the idea that natural selection explains what we observe. In both instances, the goal is to explain why a certain *pattern of variation* exists.

Why do polar bears have thicker fur than brown bears? The reason is that polar bears live in colder climates than brown bears. In particular, natural selection favored one trait for the organisms in the one habitat and a different trait for the organisms in the other. More specifically still, there are genetic differences between the two species that account for the difference in the thickness of their coats. These genetic differences arose because of selective differences in the bears' environments.

Why do the polar bears who live near the North Pole have thicker fur than the polar bears who live farther south? The reason is that the bears in the first group live in a colder climate than do the bears in the second. In particular, natural selection favored polar bears who possessed *phenotypic plasticity.* The bears evolved a set of genes that permitted them to respond to changes in the ambient temperature. The bears in the two groups differ in coat thickness, but this is not because they are genetically different. Rather, the difference in phenotype is to be explained by appeal to an environmental difference.

There is a truism that neither of these patterns of explanation contradicts. Every trait a bear has is a product of the genes it possesses interacting with the environment in which it lives. This remark is a truism because "environment" is defined as a garbage can category; it includes all factors that are not considered "genetic."

If genetic and environmental causes both play a role in the development of a trait, we still may wish to say which was more "important." This question quickly leads to nonsense if we apply it to the trait exhibited by a single organism. It is meaningless to say that 3 inches of Smokey's fur was provided by his genes and 2 inches by his environment. If we wish to assess the relative importance of causes, we must examine a population of bears in which there is variation. And, as we have seen, some patterns of variation will mainly be due to genetic differences, while others will mainly be due to differences in environment.

When sociobiologists attempt to provide evolutionary explanations of human behavioral and psychological characteristics, they are often accused of endorsing the doctrine of "genetic determinism." However, I hope these two examples show that evolutionary explanations are not automatically committed to the primacy of genetic over environmental explanation. Natural selection can have the result that phenotypic differences are due to genetic differences. But natural selection also can have the consequence that phenotypic differences are explained by differences in environment.

Let us now apply this lesson to problems like (1). Just as coat thickness can vary among bears, so views about killing can vary within and among human societies. Once again, it is useful to think about the pattern of variation that needs to be explained. Contemporary societies differ from each other, and individual societies have changed their views over time. In addition, we must remember that societies are not monolithic entities; if we look carefully *within* a society, we often will discern variation in opinion about when killing is permissible.

It is important not to confuse the code followed in a society with the simple slogans that the society endorses. First, there is the gap between ideals and reality. But, in addition, the slogans often do not begin to capture what the ideals really are. The Sixth Commandment says "Thou shalt not kill," but very few believers have thought that killing is *always* impermissible.

The ethical code endorsed by a society, or by an individual, is a complex object, one that is often difficult to articulate precisely. Ethical codes are in this respect like languages. You and I know how to speak English, but which of us is able to describe precisely the rules that define grammatical English? If we wish to explain why people have the views they do about the permissibility of killing, we first must obtain an accurate description of what those views actually are. "Why do people think that killing is wrong?" is a poorly formulated question.

Social scientists and evolutionary biologists have different contributions to make to our understanding of problem (1). In many parts of the world, capital punishment is now much less popular than it was a hundred years ago. Why did this transition occur? In contrast, consider the fact that all (or virtually all) human societies have regarded the killing of one's children as a much more serious matter than the killing of a chicken. Why is this so? Perhaps historians have more to tell us than evolutionists about the first question, but the reverse is true with respect to the second. There need be no battle between these disciplines concerning which holds the key. Ethical beliefs about killing are complex and multifaceted; different beliefs in this complex may fall within the domains of different disciplines.

If recent changes in view as to the permissibility of capital punishment don't have an evolutionary explanation, what does this mean? Of course, it is not to be denied that having ethical opinions requires a reasonably big brain, and our big brain is the product of evolution. What I mean is that the *difference* between people a hundred years ago who favored capital punishment and people now who oppose capital punishment is not to be explained by appeal to evolutionary factors. Again, the problem that matters concerns how one should explain a pattern of variation.

My example about the chicken may suggest that evolutionary considerations can be brought to bear only on characteristics that are invariant across cultures. Many sociobiologists reject this limitation. For example, Alexander (1987) and Wilson (1978) grant that human beings have enormous behavioral plasticity. We have an astonishing ability to modify our behavior in the light of environmental circumstances. Nonetheless, these sociobiologists also maintain that evolution has caused us to produce behaviors that, in the environment at hand, maximize fitness. For them, human behavioral variation is to be explained by the same kind of account I described above for geographical variation in polar bear coat thickness.

An example of this sort is provided by Alexander's (1987) explanation of the kinship system known as the avunculate. In this arrangement, a husband takes care of his sister's children far more than he takes care of his wife's. Alexander's hypothesis was that the avunculate occurs when and where it does as an adaptive response by men to high levels of female promiscuity. If a husband is more likely to be genetically related to his sister's children than to his wife's, he maximizes his inclusive fitness by directing care to his nephews and nieces.

I am not interested here in exploring whether this suggested explanation of the avunculate is empirically plausible. My point has to do with the form of the explanation proposed. Alexander is trying to explain a behavior that is *not* a cultural universal. And his preferred explanation is *not* to say that societies with the avunculate are genetically different from societies with other kinship systems. Rather, Alexander's hypothesis is that there is a universal tendency to behave in fitness-maximizing ways; this universal tendency leads people in different environments to behave differently.

My pair of examples about coat thickness in bears was intended to demonstrate that we should not equate "evolutionary explanation" with "genetic explanation." The evolutionary process can have the result that phenotypic differences are explained by genetic differences, but it also can have the result that phenotypic differences are explained by environmental differences. This flexibility in evolutionary explanation raises a question, however. If evolutionary explanations can encompass both genetic and environmental factors, what would a *non*evolutionary explanation look like?

A useful example is provided by the rapid decline in birth rates that occurred in Europe in the second half of the nineteenth century. In some areas of Italy, for instance, average family size declined from more than five children to a little more than two (Cavalli-Sforza and Feldman 1981). Why did this demographic transition occur?

Reducing family size did not enhance a parent's biological fitness. In fact, exactly the reverse was true. It is important to remember that fitness concerns survival *and* reproductive success. In this instance, forces of cultural change worked in opposition to the direction of natural selection. Cultural change overwhelmed the weaker opposing force of biological evolution. It is the historian, not the evolutionist, who will explain the demographic transition.

In discussing the fur thickness of bears, I mentioned a truism: Every trait a bear has is due to the genes it possesses interacting with the environment in which it lives. This truism applies to human beings with equal force. When a women in nineteenth-century Italy had only two children, this behavior was the product of a developmental process in which her genes and her environment both participated. However, I hope it is clear that this truism does not tell us what sort of explanation will be relevant to the explanation of *patterns of variation.* The demographic transition may have a quite different type of explanation from the account that Alexander recommends for understanding the distribution of the avunculate.

The question "Can evolution explain ethics?" has usually elicited two simple responses. The first is "*yes,* since ethics is a facet of human behavior and human behavior is a product of evolution." The second is "*no,* since ethics is a facet of human behavior and human behavior is a product of culture." Both of these sweeping pronouncements neglect that fact that "ethics" is not the name of a single simple trait, but names something complex and multifaceted. *Ethics* is a "supertrait," which has many "subtrait" components. The whole must be broken into its parts if we are to make headway on problems of explanation. It is reasonable to expect that some subtraits will have an evolutionary explanation while others will be explained culturally. For each subtrait that we wish to study, we must determine which form of explanation is more plausible.

3. THREE META-ETHICAL POSITIONS

In my opinion, it is *obvious* that biology and the social sciences can collectively claim problem (1) as their own. The alternative is to think that human behavior is not susceptible to scientific explanation at all. No argument for this negative thesis has ever been remotely plausible. And progress in biology and the social sciences lends empirical support to the idea that science does not stop where human behavior begins.

Even if the status of the explanatory problem (1) is reasonably clear, the question of justification (2) is not nearly so straightforward. Question (2) concerns a *normative* matter. Does it have an answer, and if it does, how is that answer related to nonnormative matters of fact?

First, some terminology. Let us say that a statement describes something *subjective* if its truth or falsity is settled by what some subject believes; a statement describes something *objective,* on the other hand, if its truth or falsity is independent of what anyone believes. "People believe that the Rockies are in North America" describes something subjective. "The Rockies are in North America," on the other hand, describes something objective. When people study geography, there is both a subjective and an objective side to this activity; there are the opinions that people have about geography, but in addition, there are objective geographical facts.

Many people now believe that slavery is wrong. This claim about present-day opinion describes a subjective matter. Is there, in addition to this widespread belief, a fact of the matter as to whether slavery really is wrong? *Ethical subjectivism,* as I will use the term, maintains that there are no objective facts in ethics. In ethics, there is opinion and nothing else.

According to subjectivism, neither of the following statements is true:

Killing is always wrong.
Killing is sometimes permissible.

Naively, it might seem that one or the other of these statements must be true. Subjectivists disagree. According to them, no normative ethical statement is true.

Ethical realism is a position that conflicts with ethical subjectivism. Realism says that in ethics there are facts as well as opinions. Besides the way a murder makes people feel, there is, in addition, the question of whether the action really is wrong. Realism does not maintain that it is always obvious which actions are right and which are wrong; realists realize that uncertainty and disagreement surround many ethical issues. However, for the realist, there are truths in ethics that are independent of anyone's opinion.[1]

There is a third position that bears mentioning. The position I'll call *ethical conventionalism* asserts that some ethical statements are true, but maintains that they are made true by some person's or persons' belief that they are true. For example, so-called *ethical relativists* say that murder is right or wrong only because various people in a society have come to believe that it is. And advocates of the *divine command theory* say that murder is wrong only because God believes (or says) that it is. Some versions of *existentialism* maintain that what is right or wrong for a person to do is settled by a decision the person makes about what sort of person he or she wishes to be.

This triplet of positions is summarized in Figure 5.1. They constitute the three possible answers that there can be to two sequential yes/no questions. Although it may not be clear which of these three positions is most plausible, it should be clear that they are incompatible. Exactly one of them is correct.[2]

The distinction between subjectivism and its alternatives is important to bear in mind when one considers what evolution can tell us about the status of problems like that posed by (2). Are facts about human biology going to tell us when and why murder is wrong? Or are biological facts going to show that all normative beliefs are untrue? Here we are forced to choose; evolution cannot simultaneously tell us which eth-

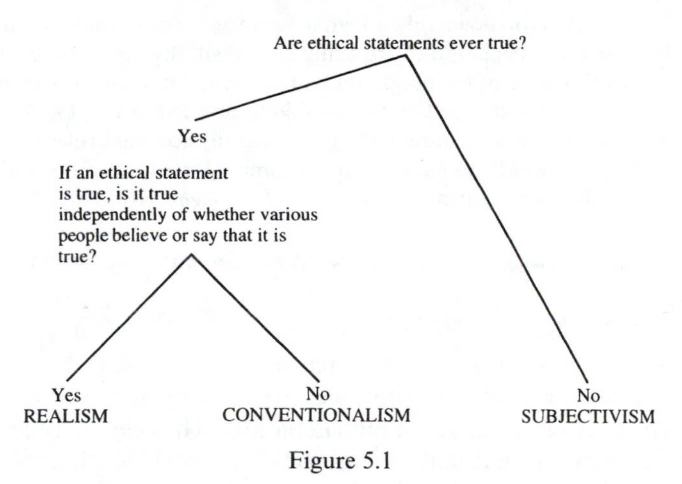
Are ethical statements ever true?
Yes
If an ethical statement is true, is it true independently of whether various people believe or say that it is true?
Yes
REALISM
No
CONVENTIONALISM
No
SUBJECTIVISM

Figure 5.1

ical norms are correct *and* that all normative statements are illusions.

Unfortunately, some evolutionary ethicists have wanted to have it both ways. For example, Ruse and Wilson (1986) have argued that an evolutionary understanding of why we have the ethical beliefs we do shows that those ethical beliefs cannot be objectively correct. Here is a characteristic passage:

> Human beings function better if they are deceived by their genes into thinking that there is a disinterested objective morality binding upon them, which all should obey. We help others because it is "right" to help them and because we know that they are inwardly compelled to reciprocate in equal measure. What Darwinian evolutionary theory shows is that this sense of "right" and the corresponding sense of "wrong," feelings we take to be above individual desire and in some fashion outside biology, are in fact brought about by ultimately biological processes. (Ruse and Wilson 1986, p. 179)

Ruse and Wilson seem to favor an emotivist account of ethical statements, according to which those statements merely express the feelings of speakers, but never say anything true. Their emotivism is a form of ethical subjectivism as defined above.

At the same time, Ruse and Wilson sometimes suggest that biology provides guidance about which norms we should adopt. Consider the following:

> Human mental development has proved to be far richer and more structured and idiosyncratic than previously suspected. The constraints on this development are the sources of our strongest feelings of right and

> wrong, and they are powerful enough to serve as a foundation for ethical codes. But the articulation of enduring codes will depend upon a more detailed knowledge of the mind and human evolution than we now possess. We suggest that it will prove possible to proceed from a knowledge of the material basis of moral feeling to generally accepted rules of conduct. To do so will be to escape – not a minute too soon – from the debilitating absolute distinction between *is* and *ought.* (Ruse and Wilson 1986, p. 174)

Here biology seems to be in the business of telling us which ethical codes are correct. Biology grounds ethics rather than unmasking it.

Ruse and Wilson sometimes draw a contrast between "internal" and "external" sources of norms. Internal norms are said to be "rooted" in our biology, whereas external ones are divinely given, or hold true independently of details about human biology. They clearly believe that ethics is "internal" and that evolutionary biology shows that this is so.

It is worth noticing that the distinction between subjectivism and realism differs from the internal/external distinction. Realists can easily maintain that an action has the ethical properties it does because of facts about our biology. When you ask your dinner companion to pass the salt and he doesn't immediately comply, why would it be wrong to stab his hand with your fork? Realists are entitled to answer this question by pointing out that stabbing someone *causes pain.* They are allowed to admit that if stabbing didn't cause pain, then the action might have quite different ethical properties. Ruse and Wilson maintain that ethics cannot be objective if it is "internal." No such consequence follows.

4. THE IS / OUGHT GAP

The goal of this paper is not to reach some general assessment of the three meta-ethical positions just described. Rather, the question is whether evolutionary considerations can tell us which is most plausible.

In this section I want to formulate and criticize one argument for subjectivism that I believe has had a great deal of influence. This argument makes no specific reference to evolutionary considerations, but it is in the background of many discussions about evolutionary ethics.

The argument I'll formulate has its provenance in Hume's distinction between *is* and *ought,* a distinction to which Ruse and Wilson alluded in the passage quoted before. I will say that an *is-statement* describes what is the case without making any moral judgment about whether this situation is good or bad. An *ought-statement,* on the other hand, makes a moral judgment about the moral characteristics (rightness, wrongness, etc.) that some action or class of actions has. For example, "thousands of people are killed by handguns every year in the United States" is an *is-statement;* "it is wrong that handguns are unregulated" is an *ought-statement.*

Hume defended the thesis that *ought-statements* cannot be deduced from exclusively *is-statements.* For example, he would regard the following argument as deductively invalid:

Torturing people for fun causes great suffering.

Torturing people for fun is wrong.

The conclusion does not follow deductively from the premisses. However, if we supply an additional premiss, the argument can be made deductively valid:

Torturing people for fun causes great suffering.
It is wrong to cause great suffering.

Torturing people for fun is wrong.

Notice that this second argument, unlike the first, has an ought-statement as one of its premisses. Hume's thesis says that *a deductively valid argument for an ought-conclusion must have at least one ought-premiss.*

The term "naturalistic fallacy" is sometimes applied to any attempt to deduce *ought*-statements from exclusively *is*-premisses. The terminology is a bit misleading, since it was G. E. Moore in *Principia Ethica* (1903) who invented the idea of a "naturalistic fallacy," and Moore's idea differs in some respects from the one just described. Unfortunately, most people discussing evolutionary ethics tend to use Moore's label to name Hume's insight. I want to keep them separate. The proposition I'll call *Hume's thesis* says you can't deduce an *ought* from an *is.* Hume's thesis, I believe, is correct.[3]

Hume's thesis, by itself, does not entail subjectivism. However, it plays a role in the following argument for subjectivism:

Ought-statements cannot be deduced validly from exclusively *is*-premisses.
If *ought*-statements cannot be deduced validly from exclusively *is*-premisses, then no *ought*-statements are true.

No *ought*-statements are true.

The first premiss is Hume's thesis. The second premiss, which is needed to reach the subjectivist conclusion, is a *reductionist assumption.* It says that for an *ought*-statement to be true, it must reduce to (be deducible from) exclusively *is*-premisses.

My doubts about this argument center on the second premiss. Why should the fact that ethics cannot be deduced from purely *is*-propositions show that no ethical statements are true? Why can't ethical statements be true, though irreducible? It is important to remember that Hume's thesis concerns *deductive* arguments. Consider an analogy: Scientific theories about unobservable entities cannot be deduced from premisses that are strictly about observables, but this provides no reason to think that theories about unobservables are always untrue.

My remarks on this argument provide no positive defense of ethical realism. However, I hope they do show that one influential argument behind ethical subjectivism is not as decisive as it might seem.

5. GENETIC ARGUMENTS

I now want to consider a second argument for ethical subjectivism. It asserts that ethical beliefs cannot be true because the beliefs we have about right and wrong are merely the product of evolution. An alternative formulation of this idea would be that subjectivism must be true because our ethical views result from the socialization we experience in early life. These two ideas may be combined as follows:

(G) We believe the ethical statements we do because of our evolution and because of facts about our socialization.

No ethical statement is true.

Philosophers are often quick to dismiss arguments like (G) on the grounds that these arguments are guilty of what has come to be called *the genetic fallacy.* A genetic argument describes the genesis (origin) of a belief and attempts to extract some conclusion about the belief's truth or plausibility.

The dim view that many philosophers take of genetic arguments reflects a standard philosophical distinction between the *context of discovery* and the *context of justification.* This distinction, emphasized by the logician Gottlob Frege, was widely embraced by the positivists.

Hempel (1965) illustrates the point of this distinction by recounting the story of the chemist Kekulé, who worked on the problem of determining the structure of benzene. After a long day at the lab, Kekulé found himself gazing wearily at a fire. He hallucinated a pair of whirling snakes, which grabbed each other's tails and formed a circle. Kekulé, in a flash of creative insight, came up with the idea of the benzene ring.

The fact that Kekulé arrived at the idea of the benzene ring while hallucinating does not settle the question of whether benzene really has that structure. It is for psychologists to describe the context of discovery – the (possibly) idiosyncratic psychological processes that led Kekulé to his insight. After Kekulé came up with his idea, he was able to do experiments and muster evidence. This latter set of considerations concerns the logic of justification.

I agree that one can't *deduce* whether Kekulé's hypothesis was true just from the fact that the idea first occurred to him in a dream. The same holds true of my friend Alf; one can't *deduce* that his belief about the shape of the earth is mistaken just from the fact that he reached this belief because of his weird ideas about perfection. However, it is a mistake to overinterpret this point. I want to suggest that there can be perfectly reasonable genetic arguments. These will be *non*deductive in form.

Consider my colleague Ben, who walks into his introduction to philosophy class one day with the idea that he will decide how many people are in the room by drawing a slip of paper from an urn. In the urn are a hundred such slips, each with a different number written on it. Ben reaches in the urn, draws a slip that says "78," and announces that he believes that exactly 78 people are present.

Surely it is reasonable to conclude that Ben's belief is probably incorrect. This conclusion is justified because of the process that led Ben to his belief. If so, the following is a perfectly sensible genetic argument:

Ben decided that there are 78 people in the room by drawing the number 78 at random from an urn.

p ══════════════════════════════

It isn't true that there are 78 people in the room.

I have drawn a double line between premiss and conclusion to indicate that the argument is not supposed to be deductively valid. The p next to the double line represents the probability that the premiss confers on the conclusion. I claim that p is high in this argument.

It is quite true that one cannot *deduce* that a proposition is untrue just from a description of how someone came to believe it. After all, the fact that Ben drew the number 78 from the urn doesn't absolutely rule out the possibility that there are exactly 78 people in the room. Still, given the process by which he formed his belief, it would be something of a miracle if his belief just happened to be correct. In this case, the context of discovery *does* provide evidence as to whether a belief is true. If so, we must be careful not to conflate two quite different formulations of what the genetic fallacy is supposed to involve:

(5) Conclusions about the truth of a proposition cannot be *deduced validly* from premisses that describe why someone came to believe the proposition.

(6) Conclusions about the truth of a proposition cannot be *inferred* from premisses that describe why someone came to believe the proposition.

I think that (5) is true but (6) is false. Inference encompasses more than deductive inference. I conclude that argument (*G*) for ethical subjectivism cannot be dismissed simply with the remark that it commits "the genetic fallacy."

The genetic argument concerning Ben's belief is convincing. Why? The reason is that *what caused him to reach the belief had nothing to do with how many students were in the room.* When this *independence relation* obtains, the genetic argument shows that the belief is implausible. In contrast, when a *dependence relation* obtains, the description of the belief's genesis can lead to the conclusion that the belief is probably correct.

As an example of how a genetic argument can show that someone's belief is probably true, consider my colleague Cathy, who decided that there are 34 people in her philosophy class by carefully counting the people present. I suggest that the premiss in the following argument confers a high probability on the conclusion:

Cathy carefully counted the people in her class and consequently believed that 34 people were present.

p ================================

34 people were present in Cathy's class.

When Cathy did her methodical counting, the thing that caused her to believe that there were 34 people present was *not* independent of how many people actually were there. Because the process of belief formation was influenced in the right way by how many people were actually in the room, we are prepared to grant that a description of the context of *discovery* provides a *justification* of the resulting belief.

Let us turn now to the argument (G) for ethical subjectivism stated before. As the comparison of Ben and Cathy shows, the argument for subjectivism is incomplete. We need to add some premiss about how the process by which we arrive at our moral beliefs is related to which moral beliefs (if any) are true. The argument requires something like the following thesis:

(A) The processes that determine what moral beliefs people have are entirely independent of which moral statements (if any) are true.

This proposition, if correct, would support the following conclusion: *The moral beliefs we currently have are probably untrue.*

The first thing to notice about this conclusion is that it does *not* say that ethical subjectivism is correct. It says that our *current* moral beliefs are probably untrue, not that *all* ethical statements are untrue. Here we have an important difference between (G) and the quite legitimate genetic arguments about Ben and Cathy. It is clear that a genetic argument might support the thesis that the ethical statements we happen to believe are untrue. I do not see how it can show that no ethical statements are true.

The next thing to notice about argument (G) concerns assumption (A). To decide whether (A) is correct, we would need to describe (i) the processes that lead people to arrive at their ethical beliefs and (ii) the facts about the world, if any, that make ethical beliefs true or false. We then would have to show that (i) and (ii) are entirely independent of each other, as (A) asserts.

Argument (G) provides a very brief answer to (i) – it cites "evolution" and "socialization." However, with respect to problem (ii), the argument says nothing at all. Of course, if subjectivism were correct, there would be no such thing as ethical facts. But to *assume* that subjectivism is true in the context of this argument would be question-begging.

Because (G) says only a little about (i) and nothing at all about (ii), I suggest that it is impossible to tell from this argument whether (A) is correct. A large number of our beliefs stem either from evolution or from socialization. Mathematical beliefs are of this sort, but that doesn't show that no mathematical statements are true (Kitcher 1993). I conclude that (G) is a weak argument for ethical subjectivism.[4]

Perhaps many of our current ethical beliefs *are* confused. I am inclined to think that morality is one of the last frontiers that human knowledge can aspire to cross. Even harder than the problems of natural science is the question of how we ought to lead our lives. This question is harder for us to come to grips with because it is clouded with self-deception. Powerful impulses hinder us from staring moral issues squarely in the face. No wonder it has taken humanity so long to traverse so modest a distance. Moral beliefs generated by superstition and prejudice probably *are* untrue. Moral beliefs with this sort of pedigree deserve to be undermined by genetic arguments. However, from this critique of some elements of existing morality, one cannot conclude that subjectivism about ethics is correct.

There is a somewhat different formulation of the genetic argument for ethical subjectivism that is worth considering. Harman (1977) suggests that we do not need to postulate the existence of ethical facts to explain why we have the ethical thoughts and feelings that we do. Psychological and evolutionary considerations suffice to do the explaining. Harman draws the conclusion that it is reasonable to deny the existence of ethical facts. His argument implicitly appeals to *Ockham's razor* (the *principle of parsimony*), which says that we should deny the existence of entities and processes that are not needed to explain anything.[5] We may schematize this argument as follows:

We do not need to postulate the existence of ethical facts to explain why people have the ethical beliefs they do.
It is reasonable to postulate the existence of ethical facts only if that postulate is needed to explain why people have the ethical beliefs they do.

═══════════════════════════════

There are no ethical facts.

As before, I draw a double line between premisses and conclusion to indicate that the argument is supposed to be nondeductive in character. Ethical subjectivism is here recommended on the ground that it is more *parsimonious* than alternative theories.[6]

I think the first premiss of this argument is correct. Psychological and biological facts about the human mind suffice to explain why we have the ethical beliefs we do. For example, we can explain why someone believes that capital punishment is wrong without committing ourselves, one way or the other, on the question of whether capital punishment really is wrong.[7]

Nonetheless, I think that the second premiss is radically implausible. For consider an analogy. Imagine that you are attending a class in statistics in which the professor repeatedly advances claims concerning how people should reason when they face inference problems of various sorts. After several weeks, a student stands up and claims that the professor's normative remarks cannot be correct, on the grounds that they do not describe or explain how people actually think and behave. The professor, I take it, would be right to reply that statistics is not the same as psychology. Maybe human beings reason in normatively *in*correct ways. The goal of statistics is not to describe or explain behavior, but to change it.

Precisely the same remarks apply to ethics. Ethics is not psychology. The point of normative ethical statements is not to *describe* why we believe and act as we do, but to *guide* our thought and behavior. We should not endorse ethical subjectivism simply because psychological explanations don't require ethical premisses.[8]

6. A NONDEDUCTIVE ANALOG OF HUME'S THESIS

Hume said you can't deduce an *ought*-conclusion from purely *is*-premisses. This thesis, I have emphasized, leaves open whether purely *is*-premisses provide *non*deductive evidence for the truth of *ought*-conclusions.

On the face of it, the nondeductive connection of *ought* and *is* may seem obvious. Surely the first of the following two statements provides *some* evidence that the second is true:

(7) Action *X* will produce more pleasure and less pain than will action *Y*.
(8) You should perform action *X* rather than action *Y*.

I agree that (7) is evidence for (8), but I suggest that the two are connected in this way only because of a background assumption that we find so obvious that it perhaps escapes our notice – that pleasure is usually good and pain is usually bad. Without some such assumption, the evidential connection of (7) to (8) is severed.

In the light of examples like (7) and (8), I propose a generalization of Hume's thesis: *Purely is-premisses cannot, by themselves, provide nondeductive support for an ought-conclusion.* In my opinion, Hume's thesis about deduction also applies to nondeductive relations.[9]

My discussion of genetic arguments may seem to undermine this general thesis. After all, I claimed that facts about how people form their ethical beliefs can provide evidence concerning whether those beliefs are true. If the relevant facts about the process of belief formation were describable by purely *is*-propositions, then we could see genetic arguments as forging a nondeductive connection between *is* and *ought.* Just as Ben's belief about the attendance at his lecture is probably *false,* and

Cathy's belief about the size of her class is probably *true,* given the procedures that each followed in forming their beliefs, so we can examine how people form their ethical convictions and draw conclusions concerning whether those ethical beliefs are probably *true.*

But there is a hitch. Descriptions of the process of belief formation cannot provide information about whether the beliefs are true unless we make assumptions about the nature of those propositions and the connections they bear to the process of belief formation.

The nature of these assumptions becomes clearer if we compare the story about Alf with the one told about Ben. When I pointed out that Alf thinks that the earth is round because he believes that the earth is perfect and that roundness is the perfect shape, this did nothing to undermine our confidence that the earth is round. We simply concluded that Alf believes the right thing for the wrong reason. In contrast, the fact that Ben formed his belief about the number of students in his classroom by drawing a ball from an urn did lead us to conclude that his belief was probably untrue. Why did we react differently to these two examples?

The reason is that we have independent reason to think that the earth is round, but none of us has independent knowledge of the number of students in Ben's class. In the former case, we begin by thinking that the earth is round, and what we learn about Alf's peculiar thought processes does nothing to undermine our confidence. In the latter case, we begin by thinking that various enrollment figures are about equally probable, and the information about Ben's draw from the urn does nothing to modify that picture. So we end as we began – by thinking that an enrollment of exactly 78 students has a low probability.

A genetic argument draws a conclusion about whether the members of some class of statements *S* are true. It draws this conclusion by describing the procedures that people follow in forming their opinions concerning the propositions in *S.* My suggestion is that you can't get something from nothing: an argument concerning the status of *S* must include some premisses about the status of the members of *S.* In particular, if you make no assumptions at all about the status of *ought*-statements, no conclusion can be drawn by a genetic argument concerning the status of those *ought*-statements. This was the point of emphasizing the role of consideration (ii) in genetic arguments. I conclude that genetic arguments, if they are to provide evidence that an *ought*-statement is untrue, must make assumptions about normative matters.

My generalization of Hume's thesis, if correct, has interesting implications concerning what evolutionary biology, and science in general, can teach us about normative issues. Let us begin with one clear format in which scientific matters can help us decide what is right and what is wrong.

Consider the relationship between statements (7) and (8). According to the thesis we are considering, (7) provides evidence for the truth of (8) only if further normative assumptions are provided. Hedonistic utilitarianism is one way to bridge the gap. This theory says that one action is morally preferable to another if the first produces more pleasure and less pain than the second. Hedonistic utilitarianism provides the sort of ethical principle that allows (7) to furnish evidence that (8) is correct.

Proposition (7) is the sort of claim that the sciences (in particular, psychology) may be in a position to establish. So, if hedonistic utilitarianism were true, the sciences would provide evidence concerning whether *ought*-statements like (8) are correct. The thing to notice about this dialectic is that the sciences do not provide evidence about (8) *on their own.* Science, on its own, does not generate ethical conclusions; rather, a more accurate formulation would be that science, *when conjoined with ethical assumptions,* can generate ethical conclusions. Hedonistic utilitarianism entails that psychological facts like (7) have ethical implications. But hedonistic utilitarianism is not something that science shows us is correct.[10]

My suspicion is that evolutionary ethics will always find itself in this situation. It may turn out that evolutionary findings do sometimes help us answer normative questions, although the proof of this pudding will be entirely in the eating. Just as hedonistic utilitarianism makes it possible for psychologists to provide information that helps decide what is right and what is wrong, this and other ethical theories may provide a similar opening for evolutionary biologists. This cannot be ruled out in advance. However, evolutionary findings will be able to achieve this result only when they are informed by ethical ideas that are not themselves supplied by evolutionary theory. Evolutionary theory cannot, all by itself, tell us whether there are any ethical facts. Nor, if ethical facts exist, can evolutionary theory tell us, all by itself, what some of those facts are. For better or worse, ethics will retain a certain degree of autonomy from the natural sciences. This doesn't mean that they are mutually irrelevant, of course. But it does mean that evolutionary ethicists who try to do too much will end up doing too little.

NOTES

1. Some philosophers whom I would want to call ethical realists refuse to apply the terms "true" and "false" to normative statements, but prefer terms like "valid" or "correct." I happen to think that use of the term "true" in this context is fairly unproblematic; if murder is sometimes permissible, then it is true that murder is sometimes permissible. In any event, my use of the term "true" is something of an expository convenience. Realists maintain, for example, that murder is either sometimes permissible or that it never is, and that the permissibility of murder is not settled by anyone's ethical opinion or say-so. This remark describes an *instance* of the realist position. To state it in full generality, some term such as "true" or "correct" is useful, if not outright necessary.
2. I've omitted *ethical skepticism* as an option in this classification of positions. "Don't know" is not an answer that is incompatible with "yes" or "no."
3. Hume's thesis, as I understood it, does not deny that there are terms in natural languages that have both normative and descriptive content. Arguably, to say that someone is *rude* or *cruel* is to advance both a descriptive and a normative claim. It would be enough for my purposes if such claims can be construed as conjunctions, one conjunct of which is purely descriptive while the other is normative.
4. It is useful to represent genetic arguments in a Bayesian format. Consider the case of Cathy. Let C be the proposition that there are 34 students in her classroom. Let B be the proposition that Cathy believes C. The process by which Cathy formed her belief will have implications about the values of $\Pr(B/C)$ and $\Pr(B/-C)$. Bayes' theorem tells us what must be true for facts about the genesis of the belief to have implications about whether the belief is probably correct:

   $$\Pr(C/\mathrm{B}) = \Pr(B/C)\Pr(C) \,/\, [\Pr(B/C)\Pr(C) + \Pr(B/-C)\Pr(-C)].$$

   If $\Pr(B/C)$ is high and $\Pr(B/-C)$ is low, then further information about the value of $\Pr(C)$ will settle whether $\Pr(C/B)$ is high or low. Notice that the process *alone* (as described by the conditional probabilities $\Pr[B/C]$ and $\Pr[B/-C]$) does not fix a value for $\Pr(C/B)$; additional information about $\Pr(C)$ is needed.
5. For discussion of the status of Ockham's razor in scientific inference, see Sober (1988) and Essay 7 in this volume.
6. Ruse and Wilson (1986, pp. 186–87) advance something very much like this parsimony argument when they say that "the evolutionary explanation makes the objective morality redundant, for even if external ethical premisses did not exist, we would go on thinking about right and wrong in the way that we do. And surely, redundancy is the last predicate that an objective morality can possess."
7. In this respect, I am in agreement with Harman when he points out that

there is a disanalogy between ethical beliefs and many perceptual beliefs. Part of the explanation of why you now believe that there is a book in front of you is that there is a book in front of you.

8. It may be replied that statistics and ethics differ in the following respect: Normative statistical claims are grounded in mathematical truths, whereas nothing comparable can be said for ethical norms. I won't here try to evaluate whether this claimed disanalogy is correct. But note that it is irrelevant to the criticism I've made of Harman's argument: Subjectivism about normative claims (whether they are ethical or statistical in character) isn't well supported by the fact that those claims don't describe or explain how people actually behave.
9. The proposed generalization of Hume's thesis parallels a familiar point about confirmation in nonethical context: An observation confirms or disconfirms a theory only in the light of background assumptions. See Sober (1988) for discussion.
10. Kitcher (1993) is careful to distinguish the following two projects:

    (B) Sociobiology can teach us facts about human beings that, in conjunction with moral principles that we already accept, can be used to derive normative principles that we have not yet appreciated.

    (D) Sociobiology can lead us to revise our system of ethical principles, not simply by leading us to accept new derivative statements – as in (B) – but by teaching us new fundamental normative principles.

    I agree with Kitcher that (B) is far more plausible than (D).

REFERENCES

Alexander, R. (1987). *The Biology of Moral Systems.* Aldine DeGruyter.

Cavalli-Sforza, L., and Feldman, M. (1981). *Cultural Transmission and Evolution.* Princeton University Press.

Harman, G. (1977). *The Nature of Morality.* Oxford University Press.

Hempel, C. (1965). *Aspects of Scientific Explanation.* Free Press.

Kitcher, P. (1993). "Four Ways to 'Biologicize' Ethics." In K. Bayertz (ed.), *Evolution und Ethik.* Reclam 1993. Reprinted in E. Sober (ed.), *Conceptual Issues in Evolutionary Biology.* MIT Press, 1993, 2nd edition.

Moore, G. (1903). *Principia Ethica.* Cambridge University Press.

Ruse, M., and Wilson, E. (1986). "Moral Philosophy As Applied Science." *Philosophy 61:* 173–92. Reprinted in E. Sober (ed.), *Conceptual Issues in Evolutionary Biology.* MIT Press, 1993, 2nd edition.

Sober, E. (1988). *Reconstructing the Past: Parsimony, Evolution and Inference.* MIT Press.

Wilson, E. (1978). *On Human Nature.* Harvard University Press.

6

Contrastive empiricism

I

Despite what Hegel may have said, syntheses have not been very successful in philosophical theorizing. Typically, what happens when you combine a thesis and an antithesis is that you get a mishmash, or maybe just a contradiction. For example, in the philosophy of mathematics, formalism says that mathematical truths are true in virtue of the way we manipulate symbols. Mathematical Platonism, on the other hand, holds that mathematical statements are made true by abstract objects that exist outside of space and time. What would a synthesis of these positions look like? Marks on paper are one thing, Platonic forms another. Compromise may be a good idea in politics, but it looks like a bad one in philosophy.

With some trepidation, I propose in this paper to go against this sound advice. Realism and empiricism have always been contradictory tendencies in the philosophy of science. The view I will sketch is a synthesis, which I call Contrastive Empiricism. Realism and empiricism are incompatible, so a synthesis that merely conjoined them would be a contradiction. Rather, I propose to isolate important elements in each and show that they combine harmoniously. I will leave behind what I regard as confusions and excesses. The result, I hope, will be neither contradiction nor mishmash.

II

Empiricism is fundamentally a thesis about *experience.* It has two parts. First, there is the idea that experience is necessary. Second, there is the

This paper is dedicated to the memories of Geoffrey Joseph and Joan Kung – two colleagues and friends from whom I learned a lot. Each influenced the way my views on scientific realism have evolved. I will miss them both.

thesis that experience suffices. Necessary and sufficient for what? Usually this blank is filled in with something like: knowledge of the world outside the mind. I will set the issue of knowledge to one side and instead will focus on the idea that experience plays a certain role in providing us with justified beliefs about the external world. Never mind what the connection is between justified belief and knowledge.

These two parts of empiricism have fared quite differently in the past 200 years or so. The idea that experience is necessary has largely lapsed into a truism. No one thinks that *a priori* reflection all by itself could lead to reasonable science. Later, I'll identify a version of this necessity thesis that is more controversial.

The other thesis – that experience somehow suffices – has been slammed pretty hard, at least since Kant. Percepts without concepts are blind. Or as we like to say now, there is no such thing as an observation language that is entirely theory-neutral. Although positivists like the Carnap of the *Aufbau* tried to show that this empiricist thesis could be made plausible, it is now generally regarded as mistaken or confused.

One vague though suggestive metaphor for what empiricism has always aimed at is this: Our knowledge cannot *go beyond* experience. Pending further clarification, it is unclear exactly how this idea should be understood. But the basic thrust of this idea has also come in for criticism. The standard point is that pretty much everything we believe about the external world goes beyond experience. Even a simple everyday claim about the commonsense characteristics of the physical objects in my environment goes beyond the experiences I have had or can ever hope to have. A consistent empiricism, so this familiar line of criticism maintains, ultimately leads to a solipsism of the present moment.

How should realism be understood? There are many realisms. Realism is often described as a thesis about what truth is or as a thesis about what is true. Neither of these is the realism I will address.

Realism as a view about the nature of truth is a semantical thesis; a realist interpretation of a set of sentences will claim that those sentences are true or false independently of human thought and language. The sentences are said to describe a mind-independent reality and to depend for their truth values on it. The standard opponent of this semantical thesis has been verificationism, which either rejects the notion of truth or reinterprets it so that truth and falsity are said to depend on us in some way. This semantical issue will not concern me further. The issue between realism and empiricism that I want to examine concedes that truth is to be understood realistically.

Realism is sometimes described as a thesis about how we should interpret the best scientific theories we now have. We should regard them as true and not simply as useful predictive devices that tell us nothing about an unobservable reality. There really are genes and quarks, so this sort of realist says.

Putnam (1978) has challenged this realist position by claiming that our present theories will probably go the way of all previous theories – future science will find them to be false. He uses this inductive argument to say that we are naive if we regard current science as true. Realism of this sort, he claims, is predicated on the unscientific expectation that the future will not resemble the past.

Putnam's argument strikes me as overstated. I don't think that *all* previous scientific theories have been found to be false in every detail. Rather, historical change has preserved some elements and abandoned others. Nevertheless, I think his skepticism about labeling all our best current theories as true is well taken. A realist in my sense may decline to say that this or that present theory is true.

Realism, in the sense at issue here, is not a thesis about what truth is; nor is it a thesis about what is true. Rather, it is a thesis about the goals of science. This is the realism that Van Fraassen (1980) singled out for criticism. Science properly aims to identify true theories about the world. Realists may refuse to assert that this or that current theory is true, though they perhaps will want to say that some theories are our current best guesses as to what is true.

What would it be to reject this thesis about the proper goal of science? Empiricism, in Van Fraassen's sense, holds that the goal of science is to say which theories are empirically adequate. Roughly, empirical adequacy consists in making predictions that are borne out in experience.

How could the search for truth and the search for empirical adequacy constitute distinct goals? Consider two theories that are *empirically equivalent,* but which are not merely notational variants of each other. Though they disagree about unobservables, they have precisely the same consequences for what our experience will be like. Not only are the theories both consistent with all the observations actually obtained to date; in addition they do not disagree about any possible observation. A realist will think that it is an appropriate scientific question to ask which theory is true; an empiricist will deny that science can or should decide this question.

The idea of empirical equivalence has had a long history. Descartes's evil-demon hypothesis was constructed to be empirically equivalent with what I'll call "normal" hypotheses describing the physical constitution of the world outside the mind. In the last hundred years, the idea of empirical equivalence has played a central role in the philosophy of physics. Mach, Poincaré, Reichenbach, and their intellectual heirs have used this idea to press foundational questions about the geometry of space and about the existence of absolute space.

In problems of this sort, realists appeal to criteria that discriminate between the two competing hypotheses and claim that those criteria are scientifically legitimate. Perhaps we should reject the evil-demon hypothesis and accept the normal hypothesis because the latter is more parsimonious, or because the former postulates the existence of an unverifiable entity. The same has been said about the existence of absolute space and the existence of universal forces, which Reichenbach (1958) introduced to play the role of an evil demon in the problem of geometric conventionalism. Realists argue that criteria of this sort provide legitimate grounds for claiming that some theories are true and others are false. Empiricists disagree, arguing that these criteria are merely "aesthetic" or "pragmatic" and should not be taken as a ground for attributing different truth values.[1]

I mention parsimony and verifiability simply as examples. Realists may choose to describe the criteria they wish to invoke in an entirely different way. The point is that realists claim that scientific reasoning is *powerful* in a way that empiricists deny.[2]

III

The development of empiricism has been guided by the following conditional: If our knowledge cannot go beyond experience, then it should be possible to delimit (i) a set of *propositions* that can be known and (ii) a set of *methods* that are legitimate for inferring what is true. For this reason empiricists have felt compelled (i) to draw a distinction between observation statements and theoretical ones and (ii) to develop a picture of the scientific method whereby the truth of theoretical statements can never be inferred from a set of observational premises.

The result is generally thought to have been a two-part disaster. The theory/observation distinction has been drawn in different ways. But each of them, I think, has either been too vague to be useful, or, if clear, has been epistemologically arbitrary. Maxwell (1962) and Hempel (1965), among many others, asked why the size of an object should determine whether it is possible to obtain reasonable knowledge about it. Apple seeds are observable by the naked eye, but genes are not. Hempel asked "So what?" – a question that empiricism's critics have continued to press.

Empiricist theories of inference have fared no better. If empiricists are to block theoretical conclusions from being drawn from observational premisses, they must narrowly limit the rules of inference that science is permitted to use. Deduction receives the empiricist seal of approval, and maybe restricted forms of "simple induction" will do so as well. But there are scientific arguments from observational premisses to observational conclusions that do not conform to such narrow strictures. Rather, they seem to require something philosophers have liked to call "abduction" – inference to the best explanation.[3] However, once these are admitted to the empiricist's organon of methods, the empiricist's strictures dissolve. The point is that inference to the best explanation also seems to allow theoretical conclusions to be drawn from observational premisses. This now-standard argument against empiricism recurs so often that it deserves a name; I call it *the garden-path argument.*

These familiar problems affect Van Fraassen's (1980) constructive empiricism just as much as they plague earlier empiricisms. Van Fraassen says that it is appropriate for science to reach a verdict on the truth value of statements that are strictly about observable entities. But when a statement says something about unobservable entities, no conclusion about its truth value should be drawn. In this case, the scientist should consider only whether the statement is empirically adequate.

Van Fraassen takes various facts about our biology to delimit what sorts of entities are observable. Observable means observable *by us.* But the question then arises as to why no legitimate forms of scientific inference can take us from premises about observables to conclusions about unobservables. As with previous empiricisms, constructive empiricism seems to impose an arbitrary limit on the kinds of inferences it deems legitimate.

There is an additional problem with Van Fraassen's approach. It concerns the concept of aboutness. The appropriate scientific attitude we should take to a statement is said to depend on what that statement is about. But what is aboutness? I see no reason to deny that the statement "All apples are green" is about everything in the universe; it says that every object is green if it is an apple. If the universe contains unobservable entities, then the generalization is about them as well. Pending some alternative interpretation of "aboutness," constructive empiricism seems to say that science should not form opinions about the truth value of any generalization (Musgrave 1985, 208; Sober 1985).

Modern empiricism has frequently been plagued by semantical problems. Carnap tried to divide theoretical from observational statements by a verificationist theory of meaning. Van Fraassen abandons this empiricist semantics, but his theory is undermined by a semantical difficulty all the same. It is aboutness, not verificationism, that causes the problem.

Realism appears strongest when it deploys criticisms of empiricism of the kinds just mentioned. The best defense is a good offense. But when one looks at the positive arguments that realists have advanced, their position looks more vulnerable. Indeed, the problems become most glaring when the issue of empirically equivalent theories is brought to the fore.

Before Putnam lapsed from the realist straight and narrow, he sketched an argument for realism that struck many philosophers as very powerful. It is encapsulated in Putnam's (1975) remark that "realism is the only philosophy that doesn't make the success of science a miracle." I now want to consider this miracle argument for realism.

The idea is this. Suppose a theory *T* is quite accurate in the predictions it generates. This is something on which the realist and the empiricist can agree. The question is then: *Why* is the theory successful? What explains the theory's predictive accuracy?

The miracle argument seeks to show that the hypothesis that the theory is true (or approximately true) is the best explanation of why the theory is predictively accurate. If the theory postulates unobservables, then the theory's predictive success is best explained by the hypothesis that the entities postulated by the theory really exist and their characteristics are roughly as the theory says they are. Here is an example of an abductive argument that leads from observational premisses to a nonobservational conclusion. The empiricist must block this argument.

Fine (1984, 84–85) has argued that the realist cannot employ abductive arguments of this sort, since they are question begging. The issue, he says, is precisely whether abduction is legitimate. Boyd (1984, 67) rejects this criticism. He claims that scientists use abductive arguments, and so it is quite permissible for a philosopher to use abduction to defend a philosophical thesis about science.

My assessment of the miracle argument differs from both Fine's and Boyd's. I have no quarrel with philosophical abductive arguments, as long as they conform to the standards used in science. The problem with the miracle argument is not that it is abductive, but that it is a very weak abductive argument.

When scientists wish to assess the credentials of an explanatory hypothesis, a fundamental question will be: What are the alternative hypotheses that compete with the one in which you are really interested? This is the idea that theory testing is a contrastive activity. To test a theory *T* is to test it against at least one competing theory *T′*.

The miracle argument fails to specify what the set of competing hypotheses is supposed to be. The hypothesis of interest is that *T* is true or approximately true in its description of unobservables. If the problem is to choose between *T* and *T′*, where *T′* is a theory that is not predictively equivalent with *T*, then the miracle argument might make sense. That is, if the choice is between the following two conjectures, there clearly can be good scientific evidence favoring the first:

(*One*) *T* is true or approximately true.
(*Two*) *T′* is true or approximately true.

But if I vary the contrasting alternative, matters change. What scientific evidence can be offered for favoring hypothesis (*One*) over the following competitor:

(*Three*) *T* is empirically adequate, though false.

If (*Three*) were true, it would not be surprising that *T* is predictively successful.

So what becomes of the thesis that realism is the only hypothesis that doesn't make the success of science a miracle? Strictly speaking, it is false. A realist interpretation of the theory *T* is given by (*One*); if true, it would explain what we observe – that *T* is predictively successful. But the same holds of (*Three*); if it were true, that also would explain the predictive success of *T*.

Notice that my critique of the miracle argument does not proceed by artificially limiting science to a discussion of observables. Nor does it reject the legitimacy of abduction. Both *T* and *T′* may talk about unobservables; and the choice between hypotheses (*One*) and (*Two*) may be an unproblematic case of inference to the best explanation. The criticism just sketched differs from the empiricist's standard position. Rather, it is characteristic of the view that I call Contrastive Empiricism.

It may be objected that hypothesis (*Three*) is really no explanation at all of the predictive success that *T* has enjoyed. If Holmes finds a corpse outside of 221B Baker Street, the hypothesis that Moriarty is the murderer is one possible explanation. But is it an explanation to assert that Moriarty is innocent, though the crime looks just as it would have if Moriarty had done the dirty deed? Does this remark explain why the murder took place?

This question is a subtle one for the theory of explanation. Perhaps there are occasions in which saying that T is not the explanation may itself be an explanation; perhaps not. What I wish to argue is that this point is irrelevant to the issue of whether the miracle argument is successful. The question before us is first and foremost a question about *confirmation.* We want to know whether the predictive accuracy of theory T is good evidence that T is true or approximately true. The issue of explanation matters here only insofar as explanation affects confirmation. My view is that loose talk about abduction has brought these two ideas closer together than they deserve to be.

I'll use Bayes' theorem to illustrate what I have in mind. This theorem says that the probability, $\Pr(H / O)$, of an hypothesis (H) in the light of the observations (O) is a function of three other probabilities:

$$\Pr(H / O) = \Pr(O / H)\Pr(H) / Pr(O).$$

We wish to compare the probability of hypothesis (*One*) and hypothesis (*Three*), given that theory T has been predictively successful (O). (*One*) is more probable than (*Three*), in the light of this observation, precisely when:

$$\Pr(O / One)\Pr(One) > \Pr(O / Three)\Pr(Three).$$

The conditional probabilities in this last expression are called *likelihoods:* The likelihood of a hypothesis is the probability it confers on the observations. Don't confuse this quantity with the hypothesis's posterior probability, which is the probability that the observations confer on the hypothesis. So whether the above inequality is true depends on the likelihoods and the prior probabilities of hypotheses (*One*) and (*Three*).

In this Bayesian format,[4] it is likelihood that represents the ability of the hypothesis to explain the observations. The question of whether the hypothesis explains the observations is interpreted to mean: how probable are the observations, if the hypothesis is true? An hypothesis with a small likelihood says that the observations are very improbable – that it is almost a "miracle," so to speak, that they occurred. Understood in this way, hypotheses (*One*) and (*Three*) are equally explanatory, since they confer the same probability on the observations.

It may be replied, with some justice, that likelihood does not fully capture the idea of explanatory power. Indeed, there are reasons independent of the problem of comparing hypotheses (*One*) and (*Three*) for thinking this. For example, two correlated effects of a common cause may make each other quite probable; given the presence of one effect, it may be more likely to infer that the other is present than that it is absent. Yet neither of the correlates explains the other. All this may be true, but my question, then, is this: When explanatory power diverges from likelihood, why think that explanatory power is relevant to confirmation? In the present case, let us grant that (*One*) is more explanatory than (*Three*). Why is this evidence that (*One*) is more plausible than (*Three*)?

Again, I want to emphasize that my criticism does not reject the idea of inference to the best explanation. Theoretical hypotheses about unobservables – like (*One*) and (*Two*) – have likelihoods. Inference to the best explanation should take those likelihoods into account. The problem, though, is that empirically equivalent theories have identical likelihoods.

I am reluctant to take "explanatory power" as an unanalyzed primitive that conveniently has just the characteristics that realists need if they are to justify their pet discriminations between pairs of empirically equivalent theories. Perhaps a non-Bayesian confirmation theory can make good on this realist idea. I don't know of any proposal that does the trick. So I am reluctant to allow the miracle argument to proceed as a resolution of the problem of choosing between (*One*) and (*Three*).

In the Bayesian inequality stated before, there are other elements besides likelihoods. In addition, there are the prior probabilities of hypotheses (*One*) and (*Three*). If (*One*) were *a priori* more probable than (*Three*), that would help the realist, although it would be wrong to say that the empirical accuracy of theory T was doing any work. If the likelihoods are the same, then the observation is idle. Perhaps we shouldn't call this the miracle argument at all; it isn't that realism is a better explanation of what we observe. Rather, the idea now is that a realist interpretation of a theory is *a priori* more probable than the alternative.

I am at a loss to see how this idea can be parlayed into a convincing argument for realism. What do these prior probabilities mean? If they are just subjective degrees of belief that some agent happens to assign, we are simply saying that this agent favors realism before any observations have been made. This is hardly an argument for realism, since another agent could have just the opposite inclination.

If prior probabilities are to be used in an argument, it must be shown why hypothesis (*One*) *should* be assigned a higher prior than (*Three*). I know of no way of doing this, though perhaps an *a priori* argument for realism will someday be invented. Let me emphasize, however, that this is worlds away from Putnam's *a posteriori* miracle argument. That argument is defective, if explanatory power goes by likelihoods; or it is entirely unclear what the argument says, if explanatory power is to be understood in some other way.[5]

I began this section by rehearsing the standard criticism that empiricism takes an overly narrow view of the scope and limits of scientific inference. Conclusions about unobservables can be blocked only by drastically restricting inferences in a way that seems entirely artificial. But the present discussion of the miracle argument suggests that realism errs in the opposite direction. The idea of inference to the best explanation presupposed by the miracle argument licenses too much, if a roughly Bayesian idea of confirmation is used.

The empiricist wants to show that there is an important sense in which our knowledge cannot go beyond experience. The realist wants to show that our ability to know about quarks is every bit as strong as our ability to find out about tables. The empiricist idea runs into trouble when it artificially limits the power of scientific inference. The realist idea runs into trouble when it artificially inflates that power. It now is time to see how the defensible kernel of each position can be formulated as a single position, one that avoids the excesses of each.

IV

I mentioned before that theory testing is a contrastive activity. If you want to test a theory *T*, you must specify a range of alternative theories – you must say what you want to test *T against*.

There is a trivial reading of this thesis that I do not intend. To find out if *T* is plausible is simply to find out if *T* is more plausible than *not-T*. I have something more in mind: there are various contrasting alternatives that might be considered. If *T* is to be tested against *T′*, one set of observations may be pertinent; but if *T* is to be tested against *T″*, a different set of observations may be needed. By varying the contrasting alternatives, we formulate genuinely different testing problems.

An analogous point has been made about the idea of explanation.[6] To explain why a proposition *P* is true, we must explain why *P*, rather than some contrasting alternative *Q*, is true (Dretske 1973; Garfinkel 1981; Van Fraassen 1980; see Salmon 1984 for criticisms and Essay 9 for discussion). This thesis is nontrivial, since varying the contrasting proposition *Q* poses different explanatory problems.

A nice example, due to Garfinkel, concerns the bank robber Willi Sutton. A priest once asked Willi why he, Willi, robbed banks. Willi answered that that was where the money was. The priest wanted to know why Willi robbed rather than not robbing. Willi took the question to be why he robbed banks rather than candy stores.

The choice of a contrasting alternative helps delimit what sort of propositions may be inserted into an answer to a why-question. If you ask why Willi robbed banks rather than candy stores, you may include in your answer the assumption that Willi was going to rob something. However, if you ask why Willi indulged in robbing instead of avoiding that activity, you cannot include the assumption that Willi was going to rob something.

Why is it legitimate to insert the statement that Willi was going to rob something into the answer to the first question, but not into the second? Consider the question "Why *P* rather than *Q*?" I claim that statements implied by both *P* and *Q* are insertable. The question presupposes the truth of such statements, so they may be assumed in the answer. On the other hand, the implications that *P* has, but *Q* does not, cannot be inserted into an answer. It is matters such as these that are at issue, so assuming them in the answer would be question-begging.

I have described a sufficient condition for insertability and a sufficient condition for noninsertability. I will not propose a complete account by specifying a necessary and sufficient condition. The modest point of importance here is that the formulation of an explanatory why-question often excludes a statement from being inserted into an answer.

I turn now from explanation to confirmation. Instead of asking "Why *P* rather than *Q*?" I want to consider the question "Why *think P* rather than *Q*?" This is a request for evidence or for an argument showing that *P* is more plausible than *Q*. I want to claim here that confirmatory why-questions often exclude some statements from being insertable.

A statement *S* will not be insertable into an answer to such questions, if it is not possible to know that *S* is true without already knowing that *P* is more plausible than *Q*. What is requested by the question is an independent reason, not a question-begging one.

As in the case of explanation, a statement may be insertable into the answer to one confirmatory why-question without being insertable into the answer to another. One way to change a question so that an answer insertable before is no longer so, is by a procedure I'll call *absorption.* Suppose I ask why Willi led a life of crime rather than going straight. You might answer by citing Willi's tormented adolescence. But if I absorb this answer into the question, I thereby obtain a new question, which cannot be answered by your previous remark. Suppose I ask: Why did Willi have a tormented adolescence and then lead a life of crime, as opposed to having an idyllic adolescence followed by a law-abiding adulthood? The assertion that Willi had a tormented adolescence is not insertable into an answer to this new question.

It is just this strategy of absorption that philosophers have used to generate skeptical puzzles about empirically equivalent theories. The question "Why think *P* rather than *Q?*" may have *O* as an answer. But *O* cannot be inserted into an answer to a new question: "Why think that *P* and *O* are true, rather than *Q* and *not-O?*" Reichenbach (1958) argued that if I assume a normal physics devoid of universal forces, I can develop experimental evidence for thinking that space is non-Euclidean rather than Euclidean. But if I absorb the assumptions about a normal physics into my question, I obtain a new question that is not, according to Reichenbach, empirically decidable; the reason, he claimed, is that the conjunction of the non-Euclidean hypothesis and normal physics is empirically equivalent to the conjunction of the Euclidean hypothesis and a physics that postulates universal forces.

As noted earlier, empiricists have always maintained that it is not possible to say that one theory has a better claim to be regarded as true than another, if the two are empirically equivalent. Such discriminatory why-questions, the empiricist claims, are unanswerable.

Empiricists have defended this view by claiming that there is a privileged set of statements – formulated in the so-called observation language. Sentences in this special class were supposed to have the following feature – ascertaining their truth required no theoretical information whatever. The empiricist claim then was made that a discrimination can be made between two theories only if they make different predictions about what will be true in this class of sentences.

The standard criticism of this idea was developed by claiming that the distinction between theory and observation is not absolute. The very statements that on some occasions provide independent answers to confirmatory why-questions at other times provide only question-begging answers. A statement that counts as an observation statement in one context can become the hypothesis under test in another. The observation/theoretical distinction, so this criticism of empiricism maintained, is context relative and pragmatic.

The standard empiricist claim about observation has a quantifier order worth noting:

(EA) There exists a set of observation statements, such that, for any two theories T and T', if it is possible to say that T is more plausible than T', then this will be because T and T' make incompatible predictions as to which members of that set are true.

A weaker thesis, which avoids an absolute distinction between theory and observation, has a different quantifier order:

(AE) For any two theories T and T', if it is possible to say that T is more plausible than T', then this will be because there exists a set of observation statements such that T and T' make incompatible predictions as to which members of that set are true.

(EA) is committed to an absolute distinction between theory and observation; the required distinction is absolute because what counts as an observation statement is invariant over the set of testing problems. (AE) is not committed to this thesis, because it is compatible with the idea that what counts as an observation is relative to the testing problem considered. Contrastive Empiricism maintains that (AE), rather than the stronger thesis (EA), is correct. What counts as an observation in a given test situation should provide non-question-begging evidence for discriminating between the competing hypotheses.

Contrastive Empiricism makes use of the concept of an observation, as does the very formulation of the problem of empirically equivalent theories, which, recall, is a problem that both realists and empiricists want to solve. I have already mentioned that what counts as an observation may vary from one testing problem to another. But more must be said about what an observation is. I won't attempt to fully clarify this concept here, but again, will content myself with a sufficient condition for empirical equivalence, one with a Quinean cast. Two theories are empirically equivalent if the one predicts the same physical stimulations to an agent's sensory surfaces as the other one does. Observational equivalence is vouchsafed by identity of sensory input.

Both (EA) and (AE) mark the special role of experience in terms of a partition of propositions. The scientist testing a pair of theories is supposed to be able to identify a class of sentences in which the so-called observation reports can be formulated. But the empiricist's point about empirically equivalent hypotheses can be made in a quite different way. Consider an analogy: When your telephone rings, that is evidence that someone has dialed your number. But the ringing of the phone when I dial your number is physically indistinguishable from the ringing that would occur if anyone else did the dialing. This is an empirical truth that can be substantiated by investigating the physical channel. The proximal state fails to uniquely determine its distal cause.[7] I don't need to invoke a special class of protocol statements and claim that they have some special epistemological status to make this simple point. Still less does the *telephone* need to be able to isolate a special class of sentences in which it can record its own physical state.

The idea of empirically equivalent hypotheses is parallel, though, of course, more general. It is basically the idea that the proximal state of the whole sentient organism, both now and in the future, fails to uniquely determine its distal cause. Whether two hypotheses are empirically equivalent is a question about the sensory channels by which distal causes can have proximal experiential effects. What engineers can tell us about telephones, psychologists will eventually be able to tell us about human beings. I don't think that the idea of empirical equivalence requires any untenable dualisms.[8]

The main departure of this "engineering" approach to the concept of empirical equivalence from earlier "linguistic" formulations is this: In the earlier version, the scientist is viewed as thinking about the world by deploying a certain *language*. The idea of empirical equivalence is then introduced by identifying a set of sentences within that very language; two theories are then said to be empirically equivalent if they make the same predictions concerning the truth of sentences in the privileged class. In the engineering version, we can talk about two theories being empirically equivalent for a given organism (or device) without supposing that the theories are formulable within the organism's language and without supposing that the organism has a language within which the experiential content of the observation is represented without theoretical contamination.[9] It's the sensory state of the organism that matters for the engineering concept, not some special class of statements that the organism formulates.

It is not to be denied that the theories that scientists standardly wish to test do not, by themselves, imply anything about the observations they will make. If neither of two theories has observational implications, then it is only in an uninteresting and vacuous sense that they are empirically equivalent. But what cannot be said of the part can be said of the whole. I take it that two, perhaps large, conjunctions of theoretical claims (including what philosophers like to call auxiliary assumptions) can have observational implications. And what is more, it sometimes can happen that two largish conjunctions can be empirically equivalent. This, I think, is what Descartes wanted to consider when he formulated his evil-demon hypothesis and what Reichenbach had in mind by his conjunction of a physics of universal forces and a geometry. Maybe these hypotheses were short on details, but I do not doubt that there are pairs of empirically equivalent theories. It is about such pairs that empiricism and realism disagree.[10]

The main departure that Contrastive Empiricism makes from previous empiricisms, including both Logical Empiricism and Constructive Empiricism, is that it is about *problems,* not *propositions.* Previous empiricisms, as I've said, have tried to discriminate one set of statements from another. Van Fraassen, like earlier empiricists, wants to say that science ought to treat some statements differently from others. Contrastive Empiricism draws no such distinction. Rather, it states that science is not in the business of discriminating between empirically equivalent hypotheses.

For example, previous empiricisms have wanted to identify a difference between the following two sentences:

(*X1*) There is a printed page before me.
(*Y1*) Space-time is curved.

I draw no such distinction between these *propositions.* Rather, my suggestion is that there is an important similarity between two *problems.* There is the problem of discriminating between (*X1*) and (*X2*). And there is the problem of discriminating between (*Y1*) and (*Y2*):

(*X2*) There is no printed page before me; rather, an evil demon makes it seem as if there is one there.
(*Y2*) Space-time is not curved; rather, a universal force makes it seem as if it is curved.

According to Contrastive Empiricism, neither of these problems (when formulated with due care) is soluble.

Although Contrastive Empiricism embodies one part of the empiricist view that knowledge cannot go beyond experience, there is nonetheless an important realist element in this view. Hypotheses about the curvature of space-time may be as testable as hypotheses about one's familiar everyday surroundings. (*X1*) can be distinguished from a variety of empirically nonequivalent alternatives, by familiar sensory means. (*Y1*) can be distinguished from a variety of empirically nonequivalent alternatives, by more recondite, though no less legitimate, theoretical means.

Contrastive Empiricism gives abduction its due. But when the explanations under test are empirically equivalent, it concedes that no difference in likelihood will be found. If we use a rough Bayesian format and claim that there is nonetheless a difference in plausibility between (*X1*) and (*X2*), or between (*Y1*) and (*Y2*), we therefore must be willing to say that there is a difference in priors. But where could this difference come from? Contrastive Empiricism claims that no such difference can be defended.

In less philosophically weighty problems of Bayesian inference, two hypotheses may have identical likelihoods, but differ in their prior probabilities for reasons that can be defended by appeal to experience. To use an old standby, if I sample at random from emeralds that exist in 1988 and find that each is green, then, relative to this observation, the following two hypotheses have identical likelihoods:

(*H1*) All emeralds are green.
(*H2*) All emeralds are green until the year 2000, but after that they turn blue.

In spite of this, I may have an empirical theory about minerals (developed before I examined even one emerald) that tells me that emerald color is very probably stable. This theory allows me to assign (*H1*) a higher prior than (*H2*).[11]

Contrastive Empiricism is not the truism that the likelihoods of empirically equivalent theories do not differ. Rather, it additionally claims that no defensible reason can be given for assigning empirically equivalent theories different priors. A pair of empirically equivalent hypotheses differs from the (*H1*)-(*H2*) pair in just this respect.

This is not to deny that human beings look askance at evil demons and their ilk. We do assume that they are implausible. In a sense, we assign them very low priors, so that even when their likelihoods are as high as the likelihoods of more "normal" sounding hypotheses, we still can say that normal hypotheses are more probable than bizarre evil-demon hypotheses in the light of what we observe. This is how we are, to be sure. But I cannot see a rational justification for thinking about the world in this way. I cannot see that we have any non-question-begging evidence on this issue. Maybe Hume was right that the combination of naturalism and skepticism has much to recommend it.

What does Contrastive Empiricism say about the principles that realists have liked to emphasize? Can't we appeal to simplicity and parsimony as reasons for rejecting evil demons and the like? Won't such considerations count as objective, since they also figure in more mundane hypothesis testing, where the candidates are not empirically equivalent? Simplicity seems to favor (*H1*) over (*H2*) when both are consistent with the observations; so won't simplicity also favor (*X1*) over (*X2*) and (*Y1*) over (*Y2*)? Here the "garden-path argument" threatens to undermine Contrastive Empiricism. If appeals to parsimony/simplicity are permissible when the problem is to discriminate between empirically *non*equivalent hypotheses, how can such appeals be illegitimate when the problem is to discriminate between hypotheses that are empirically equivalent?

Space does not permit me to discuss this issue very much. My view is that philosophers have hypostatized the principle of parsimony. There is no such thing. Rather, I think that when scientists appeal to parsimony, they are making specific background assumptions about the inference problem at hand. There is no abstract and general principle of parsimony that spans all scientific disciplines like some abductive analog to *modus ponens.* When scientists draw a smooth curve through data points, to use a standard example, they do not do this because smooth curves are simpler than bumpy ones; rather, their preference for curves in one class rather than another rests on specific assumptions about the kind of process they are modeling.

Let me give an example that illustrates what I have in mind. Charles Lyell defended the idea of uniformitarianism in geology. He argued for this view by claiming that a principle of uniformity was a first principle of scientific inference, and that his opponents were not being scientific. However, if you look carefully at what Lyell was doing, you will see that "uniformitarianism" was a very specific theory about the Earth's history. Considered as a substantive doctrine, it is simply not true that uniformitarianism's rivals must be in violation of any first principle of scientific inference (Rudwick 1970; Gould 1985). On the other hand, if one abstracts away from the geological subject matter in the hope of identifying a suitably presuppositionless principle of simplicity or uniformity, what one obtains is a principle that has no implications whatever about whether Lyell's theory was more plausible than the alternatives.

Other examples of this sort could be enumerated. It has recently been popular for biologists to argue that group selection hypotheses should be rejected because they are unparsimonious. Those arguments, I think, are either totally without merit, or implicitly assume that the preconditions for certain kinds of evolutionary processes are rarely satisfied in nature. If parsimony is just abstract numerology, it is meaningless; if it really joins the issue, it does so by making an empirical claim about how evolution proceeds (Sober 1984). The evolutionary problem of phylogenetic inference affords another example: It is an influential biological idea that parsimony can be justified as a principle of phylogenetic inference without requiring any substantive assumptions about how the evolutionary process proceeds. Again, I think this view is mistaken; see Sober (1988) for details.

I grant that a few examples do not a general argument make. I also grant that the three examples I have cited do not involve choosing between empirically equivalent theories. Could one grant my point that appeals to parsimony and simplicity involve contingent assumptions when the competing hypotheses are not empirically equivalent, but maintain that parsimony and simplicity are entirely presuppositionless when the choice is between empirically equivalent theories?

I find this view implausible. It also strikes me as pie-in-the-sky. Within a broadly Bayesian framework, it seems clear that prior probabilities are not obtainable *a priori.*[12] If a plausible non-Bayesian confirmation theory can be developed that says differently, I would like to see it. Although I grant that our understanding of nondeductive inference is far from complete, I simply do not believe that the kind of confirmation theory that realism requires will be forthcoming.[13]

In "Empiricism, Semantics, and Ontology," Carnap (1950) introduced a distinction between internal and external *questions,* which he spelled out by distinguishing one class of *propositions* from another. Quine (1951) and others took issue with this absolute theory/observation distinction, and the rest is history. With Carnap, I believe that the idea that there are two kinds of questions is right; but unlike Carnap, I do not think this notion requires a verificationist semantics or an absolute distinction between observational propositions and theoretical ones.

Contrastive Empiricism reconciles the realist idea that we can have knowledge about unobservables with the empiricist idea that knowledge cannot go beyond experience. The view derives its realist credentials from the fact that it imposes no restrictions on the vocabulary that may figure in testable *propositions;* but it retains an important empiricist element in its claim that science cannot solve discrimination *prob-*

lems in which experience makes no difference. Again, the chief innovation of this version of empiricism is its focus on problems, not propositions.[14]

Whether Contrastive Empiricism is more plausible than the thesis and antithesis from which it is fashioned turns on epistemological issues that I have not been able to fully address here. I hope, however, to have at least put a new contrastive why-question on the table: the debate between realism and previous empiricisms – whether of the Logical or the Constructive variety – needs to be enlarged. Detailed work on the theory of hypothesis testing will show whether Contrastive Empiricism is more plausible than the philosophical hypotheses with which it competes.

NOTES

1. Reichenbach (1938) is a classic example of this empiricist position.
2. I construe realsim and empiricism as theses about how theories should be judged for their plausibility; neither thesis is committed to the claim that scientists *accept* and *reject* the hypotheses they assess. For one view of this controversial matter, see Jeffrey (1956).
3. An example: when biologists argue that the current distribution of living things is evidence that the continents probably were in contact long ago, the argument is not an induction from a sample to a containing population. Biologists did not survey a set of similar planets and see that continental drift accompanied all or most biogeographic distributions of a certain kind, and then conclude that the biogeographic distribution observed here on Earth was probably due to continental drift as well. The inference goes from an observed effect to an unobserved cause. Although the hypothesis that the continents drifted apart is not known by "direct" observation, empiricists nonetheless count it as an observation statement; if we had been present and had waited around for long enough, we could have observed continental drift. So inferences with observational conclusions often require a mode of inference that is neither deductive nor inductive.
4. It will become clear later that this approach is Bayesian only in the sense that it uses Bayes' theorem; the more distinctively Bayesian idea that hypotheses always have prior probabilities is not part of what I have in mind here.
5. Boyd (1980; 1984) advances a form of the miracle argument in which the fact to be explained is the reliability of the scientific method, rather than, as here, the reliability of a given theory. Boyd's version might be viewed as a diachronic analog of the synchronic argument I have discussed. My view is that the diachronic argument faces basically the same difficulties as the synchronic version.

6. The importance of contrasting alternatives has also been explored in connection with the problem of defining what knowledge is; see Johnsen (1987) for discussion and references to the literature.
7. Of course, a probabilistic formulation can be given to this idea: The probability of the phone's sounding a certain way, given that I dial your number, is precisely the same as the probability of its sounding that way if someone else does the dialing.
8. Van Fraassen (1980) rightly emphasizes that whether something is observable is a matter for science, not armchair philosophy, to settle. However, Van Fraassen also claims that an entity whose detection requires instrumentation should not count as "observable"; with Wilson (1984), I find this restriction arbitrary, however much it may accord with ordinary usage (Sober 1985).
9. Skyrms (1984, 117) similarly argues that what counts as an "observation might have a precise description not at the level of the language of our conscious thought but only at the level of the language of the optic nerve."
10. Although Wilson (1984) is properly skeptical about the empirical equivalence of some theory pairs that philosophers have taken to be related in this way, he nonetheless grants that there are such pairs; he cites results due to Glymour and Malament as providing cases in point.
11. There is a more global form of inductive skepticism that blocks this way of discriminating between (*H1*) and (*H2*). If *all* empirical propositions – except those about one's current observations and memory traces – are called into question, what grounds are there for preferring (*H1*) to (*H2*)? This is the "theoretically barren" context in which Hempel (1965) posed his raven paradox. My view is that this skeptical challenge cannot be answered – only in the context of a background theory do observations have evidential meaning (Good 1967; Rosenkrantz 1977; Sober 1988).
12. But see Rosenkrantz (1977) for dissenting arguments.
13. This Bayesian appraoch to the conflict between realism and empiricism is very much in the spirit of Skyrms' (1984) pragmatic empiricism. Skyrms' main focus is on the idea of confirmation, which the Bayesian understands in terms of a comparison between the posterior and prior probabilities of a hypothesis; my focus has been on the idea of hypothesis testing, which is understood in terms of a comparison of two posterior probabilities. This difference in emphasis aside, we agree that empiricism should not be understood as a semantic thesis; nor should it claim that hypotheses about unobservables cannot be confirmed or tested.
14. Fine (1984) also proposes a compromise between realism and anti-realism, but not, I think, the one broached here. Fine sees the realist and the anti-realist as both "accepting the results of science." The realist augments this core position with a substantive theory of truth as correspondence, whereas the anti-realist goes beyond the core with a reductive analysis of truth, or in some other way. Fine's idea is to retain the core and reject both sorts of proposals for augmenting it. In contrast, the opposi-

tion between empiricism and realism described in the present paper does not concern the notion of truth. What is more, the realism and empiricism with which I am concerned do not in any univocal sense "accept the results of science," since realism claims that these results include discriminations between empirically equivalent theories, whereas the empiricist denies this.

REFERENCES

Boyd, R. 1980, Scientific Realism and Naturalistic Epistemology. In *PSA 1980,* vol. 2, eds. P. Asquith and R. Giere. East Lansing, Mich.: Philosophy of Science Association.

Boyd, R. 1984. "The Current Status of Scientific Realism." In *Scientific Realism,* ed. J. Leplin. Berkeley: University of California Press, 41–82.

Carnap, R. 1950. Empiricism, Semantics, and Ontology. *Revue Internationale de Philosophie* 4: 20–40. Reprinted in *Meaning and Necessity,* Chicago: University of Chicago Press, 1956.

Dretske, F. 1973. Contrastive Statements. *Philosophical Review 81:* 411–37.

Fine, A. 1984. "The Natural Ontological Attitude." In *Scientific Realism,* ed. J. Leplin. Berkeley: University of California Press, 83–107.

Garfinkel, A. 1981. *Forms of Explanation: Rethinking the Questions of Social Theory.* New Haven: Yale University Press.

Good, I. 1967. The White Shoe Is a Red Herring. *British Journal for the Philosophy of Science* 17: 322.

Gould, S. 1985. "False Premise, Good Science." In *The Flamingo's Smile.* New York: Norton, 126–38.

Hempel, C. 1965. *Philosophy of Natural Sciences.* Englewood Cliffs, N.J.: Prentice-Hall.

Jeffrey, R. 1956. Valuation and Acceptance of Scientific Hypotheses. *Philosophy of Science* 23: 237–46.

Johnsen, B. 1987. Relevant Alternatives and Demon Skepticism. *Journal of Philosophy 84:L* 643–52.

Maxwell, G. 1962. "The Ontological Status of Theoretical Entities." In *Minnesota Studies in the Philosophy of Science,* vol. 3, *Realism and Reason,* eds. H. Feigl and G. Maxwell, 3–27. Minneapolis: University of Minnesota Press.

Musgrave, A. 1985. "Realism versus Constructive Empiricism." In *Images of Science: Essays on Realism and Empiricism,* eds. P. Churchland and C. Hooker. Chicago: University of Chicago Press, 197–221.

Putnam, H. 1975. *Mind, Language, and Reality: Philosophical Papers,* vol. 2. Cambridge: Cambridge University Press.

Putnam, H. 1978. "What Is Realism?" In *Meaning and the Moral Sciences.* London: Routledge and Kegan Paul. Reprinted in *Scientific Realism,* ed. J. Leplin. Berkeley: University of California Press, 1984, 140–53.

Quine, W. 1951. "On Carnap's Views on Ontology." In *The Ways of Paradox and Other Essays.* New York: Random House, 1966.

Reichenbach, H. 1938. *Experience and Prediction.* Chicago: University of Chicago Press.

Reichenbach, H. 1958. *The Philosophy of Space and Time.* New York: Dover.

Rosenkrantz, R. 1977. *Inference, Method, and Decision.* Dordrecht: Reidel.

Rudwick, M. 1970. The Strategy of Lyell's *Principles of Geology. Isis 61:* 5–55.

Salmon, W. 1984. *Scientific Explanation and the Causal Structure of the World.* Princeton: Princeton University Press.

Skyrms, B. 1984. *Pragmatics and Empiricism.* New Haven: Yale University Press.

Sober, E. 1982. Realism and Independence. *Noûs 61:* 369–86.

Sober, E. 1984. *The Nature of Selection.* Cambridge: MIT Press.

Sober, E. 1985. Constructive Empiricism and the Problem of Aboutness. *British Journal for the Philosophy of Science 36:* 11–18.

Sober, E. 1988. *Reconstructing the Past: Parsimony, Evolution, and Inference.* Cambridge: MIT Press.

Van Fraassen, B. 1980. *The Scientific Image.* Oxford: Oxford University Press.

Wilson, M. 1984. "What Can Theory Tell Us About Observation?" In *Images of Science,* eds. P. Churchland and C. Hooker. Chicago: University of Chicago Press, 222–44.

7

Let's razor Ockham's razor

1. INTRODUCTION

When philosophers discuss the topic of explanation, they usually have in mind the following question: Given the beliefs one has and some proposition that one wishes to explain, which subset of the beliefs constitutes an explanation of the target proposition? That is, the philosophical 'problem of explanation' typically has bracketed the issue of how one obtains the beliefs; they are taken as given. The problem of explanation has been the problem of understanding the relation '*x* explains *y*'. Since Hempel (1965) did so much to canonize this way of thinking about explanation, it deserves to be called 'Hempel's problem'.

The broad heading for the present essay departs from this Hempelian format. I am interested in how we might justify some of the explanatory propositions in our stock of beliefs. Of course, issues of theory confirmation and acceptance are really not so distant from the topic of explanation. After all, it is standard to describe theory evaluation as the procedure of 'inference to the best explanation'. Hypotheses are accepted, at least partly, in virtue of their ability to explain. If this is right, then the epistemology of explanation is closely related to Hempel's problem.

I should say at the outset that I take the philosopher's term 'inference to the best explanation' with a grain of salt. Lots of hypotheses are accepted on the testimony of evidence even though the hypotheses could not possibly be explanatory of the evidence. We infer the future from the present; we also infer one event from another simultaneously occurring event with which the first is correlated. Yet the future does not explain the present; nor can one event explain another that occurs simultaneously with the first. Those who believe in inference to the best explanation may reply that they do not mean that inferring H from E requires that H explain E. They have in mind the looser requirement that H is inferrable from E only if adding H to one's total system of beliefs would maximize the overall explanatory coherence of that system. This global constraint, I think, is too vague to criticize; I suspect that 'explanatory coherence' is here used as a substitute for 'plausibility.' I doubt that plausibility can be reduced to the concept of explanatoriness in any meaningful way.

Another way in which philosophical talk of 'inference to the best explanation' is apt to mislead is that it suggests a gulf between the evaluation of explanatory hypotheses and the making of 'simple inductions'. Inductive inference, whether it concludes with a generalization or with a prediction about the 'next instance', often is assumed to markedly differ from postulating a hidden cause that explains one's observations. Again, I will merely note here my doubt that there are distinct rules for inductive and abductive inference.

Although I am not a card-carrying Bayesian, Bayes' theorem provides a useful vehicle for classifying the various considerations that might affect a hypothesis's plausibility. The theorem says that the probability that H has in the light of evidence E ($P[H/E]$) is a function of three quantities:

$$P(H/E) = P(E/H)P(H)/P(E).$$

This means that if one is comparing two hypotheses, H_1 and H_2, their overall plausibility (posterior probability) is influenced by two factors:

$$P(H_1/E) > P(H_2/E) \text{ iff } P(E/H_1)P(H_1) > P(E/H_2)P(H_2).$$

$P(H)$ is the prior probability of H – the probability it has before one obtains evidence E. $P(E/H)$ is termed the *likelihood* of H; the likelihood of H is not H's probability, but the probability that H confers on E.

Likelihood is often a plausible measure of explanatory power. If some hypothesis were true, how good an explanation would it provide of the evidence (E)? Let us consider this as a comparative problem: We observe E and wish to know whether one hypothesis (H_1) would explain E better than another hypothesis (H_2) would. Suppose that H_1 says that E was to be expected, while H_2 says that E was very improbable. Likelihood judges H_1 better supported than H_2; it is natural to see this judgment as reflecting one dimension of the concept of explanatory power.[1]

Hypotheses we have ample reason to believe untrue may nonetheless be explanatory. They may still have the property of being such that

IF they were true, they would account well for the observations. This judgment about antecedent plausibility the Bayesian tries to capture with the idea of prior probability.

There is little dispute about the relevance of likelihood to hypothesis evaluation; nor is there much dispute as to whether something besides the present observations can influence one's judgment about a hypothesis' overall plausibility. The main matter of contention over Bayesianism concerns whether hypotheses always have well-defined priors. The issue is whether prior probability is the right way to represent judgments about antecedent plausibility.

When the hypotheses in question describe possible outcomes of a chance process, assigning them prior probabilities is not controversial. Suppose a randomly selected human being has a red rash; we wish to say whether it is more probable that he has measles or mumps. The prior probability of a disease is just its population frequency. And the likelihoods also are clear; I can say how probable it would be for someone to have the red rash if he had measles and how probable the symptom would be if he had mumps. With these assignments of priors and likelihoods, I can calculate which posterior probability is greater.

Do not be misled by the terminology here. The prior probabilities in this example are not knowable *a priori*. The prior probability of the proposition that our subject has measles is the probability we assign to that disease when we do not know that he happens to have a red rash. The fact that he was randomly drawn from a population allows us to determine the prior probability by observing the population.

Matters change when the hypotheses in question do not describe the outcomes of chance processes. Examples include Newton's theory of gravity and Darwin's theory of evolution. A Bayesian will want to assign these prior probabilities and then describe how subsequent observations modify those initial assignments. Although likelihoods are often well-defined here, it is unclear what it would mean to talk about probabilities.

Bayesians sometimes go the subjective route and take prior probabilities to represent an agent's subjective degrees of belief in the hypotheses. Serious questions can be raised as to whether agents always have precise degrees of belief. But even if they did, the relevance of such prior probabilities to scientific inquiry would be questionable. If two agents have different priors, how are they to reach some agreement about which is more adequate? If they are to discuss the hypotheses under consideration, they must be able to anchor their probability assignments to something objective (or, at least, intersubjective).

Another Bayesian reaction to the problem of priors has been to argue that they are objectively determined by some *a priori* consideration. Carnap (1950) looked to the structure of the scientist's language as a source of logically defined probabilities. But since scientists can expand or contract their languages at will, it seems implausible that this strategy will be successful. More recently, Rosenkrantz (1977), building on the work of Jaynes (1968), has argued that prior probabilities can be assigned *a priori* by appeal to the requirement that a correct prior should be invariant under certain transformations of how the variables are defined. I will not discuss this line of argument here, except to note that I do not think it works.[2] Prior probabilities, I will assume, are not assignable *a priori.*

I am not a Bayesian, in the sense that I do not think that prior probabilities are always available. But the Bayesian biconditional stated above is nonetheless something I find useful. It is a convenient reminder that hypothesis evaluation must take account of likelihoods and also of the hypotheses' antecedent plausibility. Only sometimes will the latter concept be interpretable as a probability.

Notice that the Bayesian biconditional does not use the word 'explanation'. Explanations have likelihoods; and sometimes they even have priors. This means that they can be evaluated for their overall plausibility. But there is no *sui generis* virtue called 'explanatoriness' that affects plausibility.[3] Likewise, Bayes's theorem enshrines no distinction between induction and abduction. The hypotheses may be inductive generalizations couched in the same vocabulary as the observations; or the hypotheses may exploit a theoretical vocabulary that denotes items not mentioned in the description of the observational evidence.[4] Bayesianism explains why the expression 'inference to the best explanation' can be doubly misleading.

Not only does the Bayesian biconditional make no mention of 'explanatoriness'; it also fails to mention the other epistemic virtues that philosophers like to cite. Parsimony, generality, fecundity, familiarity – all are virtues that do not speak their names. Just as Bayesianism suggests that explanatoriness is not a *sui generis* consideration in hypothesis evaluation, it also suggests that parsimony is not a scientific end in itself. When parsimoniousness augments a hypothesis' likelihood or its prior probability, well and good. But parsimony, in and of itself, cannot make one hypothesis more plausible than another.

To this it may be objected that scientists themselves frequently appeal to parsimony to justify their choice of hypotheses. Since science is a paradigm (perhaps *the* paradigm) of rationality, the objection continues, does not this mean that a theory's parsimoniousness must contribute to its plausibility? How much of twentieth-century discussion of simplicity and parsimony has been driven by Einstein's remark in his 1905 paper that his theory renders superfluous the introduction of a luminiferous aether? Removing the principle of parsimony from the organon of scientific method threatens to deprive science of its results.

This objection misunderstands my thesis. I do not claim that parsimony never counts. I claim that when it counts, it counts because it reflects something more fundamental. In particular, I believe that philosophers have hypostatized parsimony. When a scientist uses the idea, it has meaning only because it is embedded in a very specific context of inquiry. Only because of a set of background assumptions does parsimony connect with plausibility in a particular research problem. What makes parsimony reasonable in one context therefore may have nothing in common with why it matters in another. The philosopher's mistake is to think that there is a single global principle that spans diverse scientific subject matters.

My reasons for thinking this fall into two categories. First, there is the general framework I find useful for thinking about scientific inference. Probabilities are not obtainable *a priori.* If the importance of parsimony is to be reflected in a Bayesian framework, it must be linked either with the likelihoods or with the priors of the competing hypotheses. The existence of this linkage is always a contingent matter that exists because some set of *a posteriori* propositions governs the context of inquiry. The second sort of reason has to do with how I understand the specific uses that scientists have made of the principle of parsimony. These case studies also suggest that there is no such thing as an *a priori* and subject matter invariant principle of parsimony.

The idea that parsimony is not a *sui generis* epistemic virtue is hardly new. Popper (1959) claims that simplicity reflects falsifiability. Jeffreys (1957) and Quine (1966) suggest that simplicity reflects high probability. Rosenkrantz (1977) seeks to explain the relevance of parsimony in a Bayesian framework. A timeslice of my former self argued that simplicity reduces to a kind of question-relative informativeness (Sober 1975).

What is perhaps more novel in my proposal is the idea that parsimony be understood locally, not globally. All the theories just mentioned attempt to define and justify the principle of parsimony by appeal to logical and mathematical features of the competing hypotheses.

An exclusive focus on these features of hypotheses is inevitable, if one hopes to describe the principle of parsimony as applying across entirely disjoint subject matters. If the parsimoniousness of hypotheses in physics turns on the same features that determine the parsimoniousness of hypotheses in biology, what could determine parsimoniousness besides logic and mathematics? If a justification for this globally defined concept of parsimony is to be obtained, it will come from considerations in logic and mathematics. Understanding parsimony as a global constraint on inquiry thus leads naturally to the idea that it is *a priori* justified. My local approach entails that the legitimacy of parsimony stands or falls, in a particular research context, on subject matter specific (and *a posteriori*) considerations.[5]

In what follows I will discuss two examples of how appeals to parsimony have figured in recent evolutionary biology. The first is George C. Williams' use in his landmark book *Adaptation and Natural Selection* of a parsimony argument to criticize hypotheses of group selection. The second is the use made by cladists and many other systematic biologists of a parsimony criterion to reconstruct phylogenetic relationships among taxa from facts about their similarities and differences.

Williams' (1966) parsimony argument against group selection encountered almost no opposition in the evolution community. Group selection hypotheses were said to be less parsimonious than lower-level selection hypotheses, but no one seems to have asked why the greater parsimony of the latter was any reason to accept them as true.

Cladistic parsimony, on the other hand, has been criticized and debated intensively for the last twenty years. Many biologists have asserted that this inference principle assumes that evolution proceeds parsimoniously and have hastened to add that there is ample evidence that evolution does no such thing. Cladists have replied to these criticisms and the fires continue to blaze.

My own view is that it is perfectly legitimate, in both cases, to ask why parsimony is connected with plausibility. I will try to reconstruct the kind of answer that might be given in the case of the group selection issue. I also will discuss the way this question can be investigated in the case of phylogenetic inference.

I noted earlier that the Bayesian biconditional suggests two avenues by which parsimony may impinge on plausibility. It may affect the prior probabilities and it may affect the likelihoods. The first biological example takes the first route, while the second takes the second.

2. PARSIMONY AND THE UNITS OF SELECTION CONTROVERSY

Williams' 1966 book renewed contact between two disciplinary orientations in evolutionary biology that should have been communicating, but did not, at least not very much. Since the 1930s, population geneticists – pre-eminently Fisher (1930), Haldane (1932), and Wright (1945) – had been rather sceptical of the idea that there are group adaptations. A group adaptation is a characteristic that exists because it benefits the group in which it is found. Evolutionists have used the word 'altruism' to label characteristics that are disadvantageous for the organisms possessing them, though advantageous to the group. Population geneticists generally agreed that it is very difficult to get altruistic characteristics to evolve and be retained in a population. Field naturalists, on the other hand, often thought that characteristics observed in nature are good for the group though bad for the individuals. These field naturalists paid little attention to the quantitative models that the geneticists developed. These contradictory orientations coexisted for some thirty years.

Williams (1966) elaborated the reigning orthodoxy in population genetics; but he did so in English prose, without recourse to mathematical arguments. He argued that hypotheses of group adaptation and group selection are often products of sloppy thinking. A properly rigorous Darwinism should cast the concept of group adaptation on the same rubbish heap onto which Lamarckism had earlier been discarded.

Williams deployed a variety of arguments, some better than others. One prominent argument was that group selection hypotheses are less parsimonious than hypotheses that claim that the unit of selection is the individual or the gene.

This argument begins with the observation that 'adaptation is an onerous principle', one that a scientist should invoke only if driven to it. Flying fish return to the water after sailing over the waves. Why do they do this? Williams claims that there is no need to tell an adaptationist story. The mere fact that fish are heavier than air accounts for the fact that what goes up must come down. The thought that flying fish evolved a specific adaptation for returning to the water, Williams concludes, is unparsimonious and so should be rejected.

This idea – that it is more parsimonious to think of the fish's return to the water as a 'physical inevitability' rather than as an adaptation – is only part of the way the principle of parsimony applies to evolutionary explanations. Williams invokes Lloyd Morgan's rule that lower-level explanations are preferable to higher-level ones; Williams takes this to mean that it is better to think of a characteristic as having evolved for the good of the organism possessing it than to view it as having evolved for the good of the group. The principle of parsimony generates a hierarchy: purely physical explanations are preferable to adaptationist explanations, and hypotheses positing lower-level adaptations are preferable to ones that postulate adaptations at higher levels of organization.

Before explaining in more detail what Williams had in mind about competing units of selection, a comment on his flying fish is in order. I want to suggest that parsimony is entirely irrelevant to this example. If flying fish return to the water because they are *heavier than air,* then it is fairly clear why an adaptationist story will be implausible. Natural selection requires variation. If being heavier than air were an adaptation, then some ancestral population must have included organisms that were heavier than air and ones that were lighter. Since there is ample room to doubt that this was ever the case, we can safely discard the idea that being heavier than air is an adaptation. My point is that this reasoning is grounded in a fact about how natural selection proceeds and a plausible assumption about the character of ancestral populations. There is no need to invoke parsimony to make this point; Ockham's razor can safely be razored away.

Turning now to the difference between lower-level and higher-level hypotheses of adaptation, let me give an example of how Williams' argument proceeds. Musk oxen form a circle when attacked by wolves, with the adult males on the outside facing the attack and the females and young protected in the interior. Males therefore protect females and young to which they are not related. Apparently, this characteristic is good for the group, but deleterious for the individuals possessing it. A group selection explanation would maintain that this wagon-training behaviour evolved because groups competed against other groups. Groups that wagon train go extinct less often and found more daughter colonies than groups that do not.

Williams rejected this hypothesis of group adaptation. He proposes the following alternative. In general, when a predator attacks a prey organism, the prey can either fight or flee. Selection working for the good of the organism will equip an organism with optimal behaviours, given the nature of the threatening predator. If the threat comes from a predator that is relatively small and harmless, the prey is better off standing its ground. If the threat is posed by a large and dangerous predator, the prey is better off running away. A prediction of this idea is that there are some predators that cause large prey to fight and small prey to flee. Williams proposes that wolves fall in this size range; they make the male oxen stand their ground and the females and young flee to the interior. The group characteristic of wagon-training is just a statistical consequence of each organism's doing what is in its own self-interest. No need for group selection here; the more parsimonious individual-selection story suffices to explain.

Williams' book repeatedly deploys this pattern of reasoning. He describes some characteristic found in nature and the group selection explanation that some biologist has proposed for it. Williams then suggests that the characteristic can be explained purely in terms of a lower-level selection hypothesis. Rather than suspending judgment about which explanation is more plausible, Williams opts for the lower-level story, on the grounds that it is more parsimonious.

Why should the greater parsimony of a lower-level selection hypothesis make that hypothesis more plausible than an explanation in terms of group selection? Williams does not address this admittedly philosophical question. I propose the following reconstruction of Williams' argument. I believe that it is the best that can be done for it; in addition, I think that it is none too bad.

Williams suggests that the hypothesis of group selection, if true, would explain the observations, and that the same is true for the hypothesis of individual selection that he invents. This means, within the format provided by the Bayesian biconditional of the previous section, that the two hypotheses have identical likelihoods. If so, the hypotheses will differ in overall plausibility only if they have different priors. Why think that it is antecedently less probable that a characteristic has evolved by group selection than that it evolved by individual selection?

As noted earlier, an altruistic characteristic is one that is bad for the organism possessing it, but good for the group in which it occurs. Here good and bad are calculated in the currency of fitness – survival and reproductive success.[6] This definition of altruism is illustrated in the fitness functions depicted in Figure 1.1 (Essay 1). An organism is better off being selfish (S) than altruistic (A), no matter what sort of group it inhabits. Let us suppose that the fitness of a group is measured by the average fitness of the organisms in the group; this is represented in the figure by $\bar{w}$. If so, groups with higher concentrations of altruists are fitter than groups with lower concentrations.

What will happen if S and A evolve within the confines of a single population? With some modest further assumptions (e.g., that the traits are heritable), we may say that the population will evolve to eliminate altruism, no matter what initial composition the population happens to have.

For altruism to evolve and be maintained by group selection, there must be variation among groups. An ensemble of populations must be postulated, each with its own local frequency of altruism. Groups in which altruism is common must do better than groups in which altruism is rare.

To make this concrete, let us suppose that a group will fission into a number of daughter colonies once it reaches a certain census size. Suppose that this critical mass is 500 individuals and that the group will then divide into 50 offspring colonies containing 10 individuals each. Groups with higher values of $\bar{w}$ will reach this fission point more quickly, and so will have more offspring; they are fitter. In addition to this rule about colonization, suppose that groups run higher risks of extinction the more saturated they are with selfishness. These two assumptions about colonization and extinction ground the idea that altruistic groups are fitter than selfish ones – they are more reproductively successful (i.e., found more colonies) and they have better chances of surviving (avoiding extinction).

So far I have described how group and individual fitnesses are related, and the mechanism by which new groups are founded. Is that enough to allow the altruistic character to evolve? No it is not. I have omitted the crucial ingredient of *time*.

Suppose we begin with a number of groups, each with its local mix of altruism and selfishness. If each group holds together for a sufficient length of time, selfishness will replace altruism within it. Each group, as Dawkins (1976) once said, is subject to 'subversion from within'. If the groups hold together for too long, altruism will disappear before the groups have a chance to reproduce. This means that altruism cannot evolve if group reproduction happens much more slowly than individual reproduction.

I have provided a sketch of how altruism can evolve by group selection. One might say that it is a 'complicated' process, but this is not why such hypotheses are implausible. Meiosis and random genetic drift also may be 'complicated' in their way, but that is no basis for supposing that they rarely occur. The rational kernel of Williams' parsimony argument is that the evolution of altruism by group selection requires a number of restrictive assumptions about population structure. Not only must there be sufficient variation among groups, but rates of colonization and extinction must be sufficiently high. Other conditions are required as well. This coincidence of factors is not impossible; indeed, Williams concedes that at least one well documented case has been found (the evolution of the *t*-allele in the house mouse). Williams' parsimony argument is at bottom the thesis that natural systems rarely exemplify the suite of biological properties needed for altruism to evolve by group selection.[7]

Returning to the Bayesian biconditional, we may take Williams to be saying that the prior probability of a group selection hypothesis is lower than the prior probability of an hypothesis of individual selection. Think of the biologist as selecting at random a characteristic found in some natural population (like musk oxen wagon-training). Some characteristics may have evolved by group selection, others by lower-level varieties of selection. In assigning a lower prior probability to the group selection hypothesis, Williams is making a biological judgment about the relative frequency of certain population structures in nature.

In the ten years following Williams' book, a number of evolutionists investigated the question of group selection from a theoretical point of view. That is, they did not go to nature searching for altruistic characteristics; rather, they invented mathematical models for describing how altruism might evolve. The goal was to discover the range of parameter values within which an altruistic character can increase in frequency and then be maintained. These inquiries, critically reviewed in Wade (1978), uniformly concluded that altruism can evolve only within a narrow range of parameter values. The word 'parsimony' is not prominent in this series of investigations; but these biologists, I believe, were fleshing out the parsimony argument that Williams had earlier constructed.

If one accepts Williams' picture of the relative frequency of conditions favourable for the evolution of altruism, it is quite reasonable to assign group selection explanations a low prior probability. But this assignment cuts no ice, once a natural system is observed to exhibit the population structure required for altruism to evolve. Wilson (1980) and others have argued that such conditions are exhibited in numerous species of insects. Seeing Williams' parsimony argument as an argument about prior probabilities helps explain why the argument is relevant *prima facie,* though it does not prejudge the upshot of more detailed investigations of specific natural systems.

Almost no one any longer believes the principle of indifference (a.k.a. the principle of insufficient reason). This principle says that if P_1, P_2, . . . P_n are exclusive and exhaustive propositions, and you have no more reason to think one of them true than you have for any of the others, you should assign them equal probabilities. The principle quickly leads to contradiction, since the space of alternatives can be partitioned in different ways. The familiar lesson is that probabilities cannot be extracted from ignorance alone, but require substantive assumptions about the world.

It is interesting to note how this standard philosophical idea conflicts with a common conception of how parsimony functions in hypothesis evaluation. The thought is that parsimony considerations allow us to assign prior probabilities and that the use of parsimony is 'purely methodological', presupposing nothing substantive about the way the world is. The resolution of this contradiction comes with realizing that whenever parsimony considerations generate prior probabilities for competing hypotheses,[8] the use of parsimony cannot be purely methodological.

3. PARSIMONY AND PHYLOGENETIC INFERENCE

The usual philosophical picture of how parsimony impinges on hypothesis evaluation is of several hypotheses that are each consistent with the evidence,[9] or explain it equally well, or are equally supported by it. Parsimony is then invoked as a further consideration. The example I will now discuss – the use of a parsimony criterion in phylogenetic inference – is a useful corrective to this limited view. In this instance, parsimony considerations are said to affect how well supported a hypothesis is by the data. In terms of the Bayesian biconditional, parsimony is relevant because of its impact on likelihoods, not because it affects priors.

Although parsimony considerations arguably have been implicit in much work that seeks to infer phylogenetic relationships among species from facts concerning their similarity and differences, it was not until the 1960s that the principle was explicitly formulated. Edwards and Cavalli-Sforza (1963, 1964), two statistically minded evolutionists, put it this way: 'the most plausible estimate of the evolutionary tree is that which invokes the minimum net amount of evolution'. They claim that the principle has intuitive appeal, but concede that its presuppositions are none too clear. At about the same time, Willi Hennig's (1966) book appeared in English; this translated an expanded version of his German work of 1950. Although Hennig never used the word 'parsimony', his claims concerning how similarities and differences among taxa provide evidence about their phylogenetic relationships are basically equivalent to the parsimony idea. Hennig's followers, who came to be called 'cladists,' used the term 'parsimony' and became that concept's principal champions in systematics.

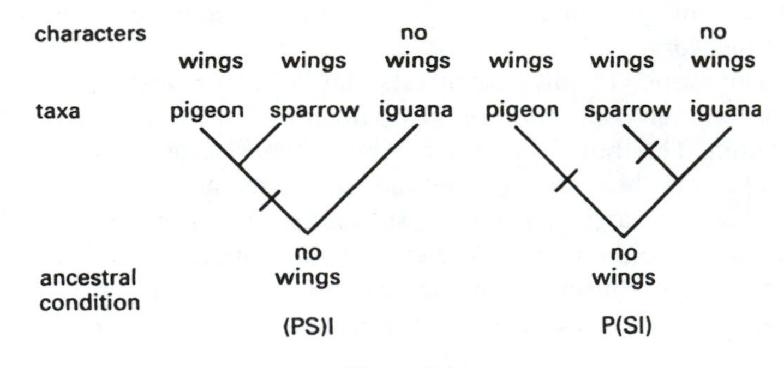

Figure 7.1

Sparrows and pigeons have wings, whereas iguanas do not. That fact about similarity and difference seems to provide evidence that sparrows and pigeons are more closely related to each other than either is to iguanas. But why should this be so?

Figure 7.1 represents two phylogenetic hypotheses and the distribution of characters that needs to be explained. Each hypothesis says that two of the taxa are more closely related to each other than either is to the third. The inference problem I want to explore involves two evolutionary assumptions. Let us assume that the three taxa, if we trace them back far enough, share a common ancestor. In addition, let us assume that this ancestor did not have wings. That is, I am supposing that having wings is the derived (apomorphic) condition and lacking wings is the ancestral (plesiomorphic) state.[10]

According to each tree, the common ancestor lacked wings. Then, in the course of the branching process, the character must have changed to yield the distribution displayed at the tips. What is the minimum number of changes that would allow the (PS)I tree to generate this distribution? The answer is *one.* The (PS)I tree is consistent with the supposition that pigeons and sparrows obtained their wings from a common ancestor; the similarity might be a homology. The idea that there was a single evolutionary change is represented in the (PS)I tree by a single slash mark across the relevant branch.

Matters are different when we consider the P(SI) hypothesis. For this phylogenetic tree to generate the data found at the tips, at least two changes in character state are needed. I have drawn two slash marks in the P(SI) tree to indicate where these might have occurred. According to this tree, the similarity between pigeons and sparrows cannot be a homology, but must be the result of independent origination. The term for this is 'homoplasy'.

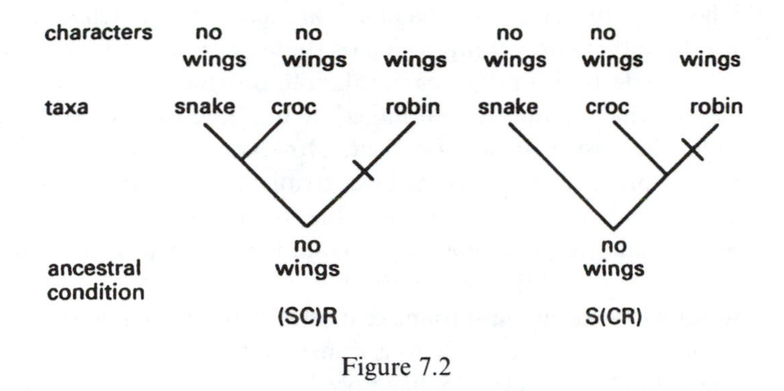

Figure 7.2

The principle of parsimony judges that the character distribution just mentioned supports (PS)I better than it supports P(SI). The reason is that the latter hypothesis requires at least two evolutionary changes to explain the data, whereas the former requires only one. The principle of parsimony says that we should minimize assumptions of homoplasy.

The similarity uniting pigeons and sparrows in this example is a *derived* similarity. Pigeons and sparrows are assumed to share a characteristic that was *not* present in the common ancestor of the three taxa under consideration. This fact about the example is important, because the principle of phylogenetic parsimony entails that some similarities do *not* count as evidence of common ancestry. When two taxa share an *ancestral* character, parsimony judges that fact to be devoid of evidential meaning.

To see why, consider the fact that snakes and crocodiles lack wings, whereas robins possess them. Does the principle of parsimony judge that to be evidence that snakes and crocodiles are more closely related to each other than either are to robins? The answer is *no,* because evolutionists assume that the common ancestor of the three taxa did not have wings. The (SC)R tree displayed in Figure 7.2 can account for this character distribution by assuming a single change; the same is true for the S(CR) tree.

In summary, the idea of phylogenetic parsimony boils down to two principles about evidence:

> Derived similarity *is* evidence of propinquity of descent.
> Ancestral similarity is *not* evidence of propinquity of descent.

It should be clear from this that those who believe in parsimony will reject *overall* similarity as an indicator of phylogenetic relationships.

For the last twenty years, there has been an acrimonious debate in the biological literature concerning which of these approaches is correct.

Why should derived similarities be taken to provide evidence of common ancestry, as parsimony maintains? If multiple originations were impossible, the answer would be clear. However, no one is prepared to make this process assumption. Less stringently, we can say that if evolutionary change were very improbable, then it would be clear why a hypothesis requiring two changes in state is inferior to a hypothesis requiring only one. But this assumption also is problematic. Defenders of parsimony have been loath to make it, and critics of parsimony have been quick to point out that there is considerable evidence against it. Felsenstein (1983), for example, has noted that it is quite common for the most parsimonious hypothesis obtained for a given data set to require multiple changes on a large percentage of the characters it explains.

I have mentioned two possible assumptions, each of which would *suffice* to explain why shared derived characters are evidence of common ancestry. No one has shown that either of these assumptions is *necessary,* though critics of parsimony frequently assert that parsimony assumes that evolutionary change is rare or improbable.

There is a logical point about the suggestions just considered that needs to be emphasized. A phylogenetic hypothesis, all by itself, does not tell you whether a given character distribution is to be expected or not. On the other hand, if we append the assumption that multiple change is impossible or improbable, then the different hypotheses do make different predictions about the data. By assuming that multiple changes are improbable, it is clear why one phylogenetic hypothesis does a better job of explaining a derived similarity than its competitors. The operative idea here is *likelihood:* If one hypothesis says that the data were hardly to be expected, whereas a second says that the data were to be expected, it is the second that is better supported. The logical point is that *phylogenetic hypotheses are able to say how probable the observations are only if we append further assumptions about character evolution.*

Figure 7.3 allows this point to be depicted schematically. The problem is to infer which two of the three taxa *A, B,* and *C* share a common ancestor apart from the third. Each character comes in two states, denoted by '0' or '1'. It is assumed that 0 is the ancestral form. Character I involves a derived similarity that *A* and *B* possess but *C* lacks.

The branches in each figure are labelled. With the *i*th branch, we can associate a transition probability e_i and a transition probability r_i. The

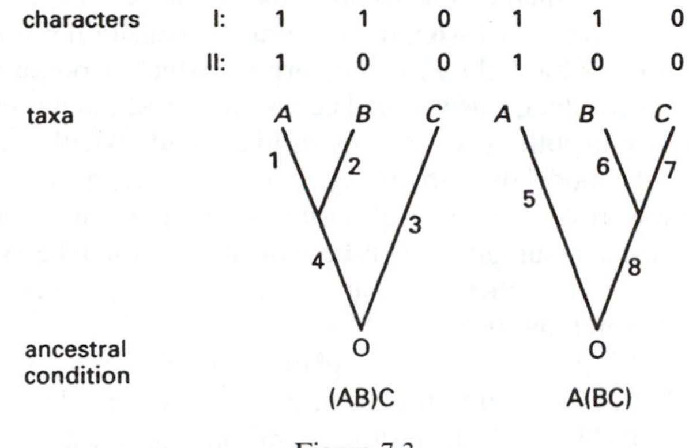

Figure 7.3

former is the probability that a branch will end in state 1, if it begins in state 0; the latter is the probability that the branch will end in state 0, if it begins in state 1. The letters are chosen as mnemonics for 'evolution' and 'reversal'.[11]

Given this array of branch transition probabilities, we can write down an expression that represents the probability of obtaining a given character distribution at the tips, conditional on each of the hypotheses. That is, we can write down an expression full of *e*'s and *r*'s that represents P[110 / (*AB*)*C*] and an expression that represents P[110 / *A*(BC)). Which of these will be larger?

If we assume that changes on all branches have a probability of 0.5, then the two hypotheses have identical likelihoods. That is, under this assumption about character evolution, the character distribution fails to discriminate between the hypotheses; it is evidentially meaningless. On the other hand, if we assume that change is very improbable on branches, 1, 2, and 4, but is not so improbable on branches 5 and 6, we will obtain the result that the first character supports (*AB*)*C less well* than it supports *A*(*BC*). This paradoxical result flies in the face of what parsimony maintains (and contradicts the dictates of overall similarity as well). As a third example, we might assign the *e* and *r* parameters associated with each branch a value of 0.1. In this case, (*AB*)*C* turns out to be more likely than *A*(*BC*), relative to the 110 distribution.

With different assumptions about branch transition probabilities, we get different likelihoods for the two hypotheses relative to character I. A similar result obtains if we ask about the evidential significance of character II, in which there is an ancestral similarity uniting *B* and *C* apart from *A*. Parsimony judges such similarities to be evidentially meaningless, but whether the likelihoods of the two hypotheses are identical depends on the assignment of branch transition probabilities.

If parsimony is the right way to determine which hypothesis is best supported by the data, this will be because the most parsimonious hypothesis is the hypothesis of maximum likelihood. Whether this is so depends on the model of character evolution one adopts. Such a model will inevitably rest on biological assumptions about evolution. One hopes that these assumptions will be plausible and maybe even commonsensical. But that they are biology, not pure logic or mathematics, is, I think, beyond question.

I will not take the space here to explain the results of my own investigation (in Sober 1988b) into the connection between phylogenetic parsimony and likelihood.[12] The more general philosophical point I want to make is this: If parsimony is the right method to use in phylogenetic inference, this will be because of specific facts about the phylogenetic process. The method does not have an *a priori* and subject matter neutral justification.

By now it is an utterly familiar philosophical point that a scientific hypothesis (H) has implications (whether deductive or probabilistic) about observations (O) only in the context of a set of auxiliary assumptions (A). Sometimes this is called Duhem's thesis; sometimes it is taken to be too obvious to be worth naming. It is wrong to think that H makes predictions about O; it is the conjunction $H\&A$ that issues in testable consequences.

It is a small step from this standard idea to the realization that quality of explanation must be a three-place relation, not a binary one. If one asks whether one hypothesis (H_1) provides a better explanation of the observations (O) than another hypothesis (H_2) does, it is wrong to think this is a matter of comparing how H_1 relates to O with how H_2 relates to O. Whether H_1 is better or worse as an explanation of O depends on further auxiliary assumptions A.

Why should a more parsimonious explanation of some observation be better than one that is less parsimonious? This cannot be a matter of the logical relationship between hypothesis and observation alone; it must crucially involve auxiliary assumptions.[13]

4. CONCLUDING REMARKS

Philosophical discussion of simplicity and parsimony is replete with remarks to the effect that smooth curves are better than bumpy ones and that postulating fewer entities or processes is better than postulating more. The natural philosophical goal of generality has encouraged the idea that these are methodological maxims that apply to all scientific subject matters. If this were so, then whatever justification such maxims possess would derive from logic and mathematics, not from anything specific to a single scientific research problem.

Respect for the results of science then leads one to assume that general principles of simplicity and parsimony must be justified. The question is where the global justification is to be found; philosophers have been quite inventive in generating interesting proposals. As a fallback position, one could announce, if such proposals fail, that simplicity and parsimony are *sui generis* constituents of the habit of mind we call 'scientific'. According to this gambit, it is part of what we mean by science that simpler and more parsimonious hypotheses are scientifically preferable. Shades of Strawson on induction.

Aristotle accused Plato of hypostatizing The Good. What do a good general and a good flute player have in common? Aristotle argued that the characteristics that make for a good military commander need have nothing in common with the traits that make for a good musician. We are misled by the common term if we think that there must be some property that both possess, in virtue of which each is good.

Williams argued that we should prefer lower-level selection hypotheses over group selection hypotheses, since the former are more parsimonious. Hennig and his followers argued that we should prefer hypotheses requiring fewer homoplasies over ones that require more, since the former are parsimonious. Following Aristotle, we should hesitate to conclude that if Williams and Hennig are right, then there must be some single property of parsimonious hypotheses in virtue of which they are good.

Maxims like Ockham's razor have their point. But their force derives from the specific context of inquiry in which they are wielded. Even if one is not a Bayesian, the Bayesian biconditional provides a useful reminder of how parsimony may affect hypothesis evaluation. Scientists may assign more parsimonious hypotheses a higher antecedent plausibility; but just as prior probabilities cannot be generated from ignorance, so assignments of prior plausibility must be justified by concrete assumptions if they are to be justified at all. Alternatively, scientists may assert that more parsimonious hypotheses provide better explanations of the data; but just as scientific hypotheses standardly possess likelihoods only because of an assumed model connecting hypothesis to data, so assessments of explanatory power also must be justified by concrete assumptions if they are to be justified at all.[14]

By suggesting that we razor Ockham's razor, am I wielding the very instrument I suggest we abandon? I think not. It is not abstract numerology – a formal preference for less over more – that motivates my conclusion. Rather, the implausibility of postulating a global criterion has two sources. First, there are the 'data'; close attention to the details of how scientific inference proceeds in well-defined contexts of inquiry suggests that parsimony and plausibility are connected only because some local background assumptions are in play. Second, there is a more general framework according to which the evidential connection between observation and hypothesis cannot be mediated by logic and mathematics alone.[15]

Admittedly, two is not a very large sample size. Perhaps, then, this paper should be understood to provide a *prima facie* argument for concluding that the justification of parsimony must be local and subject matter specific. This sort of circumspection is further encouraged by the fact that many foundational problems still remain concerning scientific inference. I used the Bayesian biconditional while eschewing Bayesianism; I offered no alternative doctrine of comparable scope. Nonetheless, I hope that I have provided some grounds for thinking that razoring the razor may make sense.

NOTES

1. Although I agree with Salmon (1984) that a true explanation can be such that the *explanans* proposition says that the *explanandum* proposition had low probability, I nonetheless think that the explanatory power of a candidate hypothesis is influenced by how probable it says the *explanandum* is. See Sober (1987) for further discussion. It also is worth noting that philosophical discussion of explanation has paid little attention to the question of what makes one explanation a better explanation than another. Hempel's problem leads one to seek a yes/no criterion for being an explanation, or for being an ideally complete explanation. It is another matter to search for criteria by which one hypothesis is a better explanation than another.
2. See Seidenfeld's (1979) review of Rosenkrantz's book for some powerful objections to objective Bayesianism. Rosenkrantz (1979) is a reply.
3. Here I find myself in agreement with Van Fraassen (1980), 22.
4. This is why a Bayesian model of theory testing counts against Van Fraassen's (1980) constructive empiricism. According to Van Fraassen, the appropriate epistemic attitude to take towards a hypothesis depends on what the hypothesis is about. If it is strictly about observables, it is a legitimate scientific task to say whether the hypothesis is true or false. If it is at least partly about unobservables, science should not pronounce on this issue.

These strictures find no expression in the Bayesian biconditional. I discuss the implications of the present view of confirmation for the realism/empiricism debate in Essay 6.

5. Miller (1987) also develops a local approach to confirmational issues, but within a framework less friendly to the usefulness of Bayesian ideas.
6. I will ignore the way the concept of inclusive fitness affects the appropriate definition of altruism and, indirectly, of group selection. I discuss this in Sober (1984, 1988c).
7. I believe that this reconstruction of Williams' parsimony argument is more adequate than the ones I suggest in Sober (1981, 1984).
8. Philosophers have sometimes discussed examples of competing hypotheses that bear implication relations to each other. Popper (1959) talks about the relative simplicity of the hypothesis that the earth has an elliptical orbit and the hypothesis that it has a circular orbit, where the latter is understood to entail the former. Similarly, Quine (1966) discusses the relative simplicity of an estimate of a parameter that includes one significant digit and a second estimate consistent with the first that includes three. In such cases, saying that one hypothesis has a higher prior than another of course requires no specific assumptions about the empirical subject at hand. However, it is debatable whether these are properly treated as competing hypotheses; and even if they could be so treated, such purely logical and mathematical arguments leave wholly untouched the more standard case in which competing hypotheses do not bear implication relations to each other.
9. Although 'consistency with the data' is often how philosophers describe the way observations can influence a hypothesis' plausibility, it is a sorry explication of that concept. For one thing, consistency is an all or nothing relationship, whereas the support of hypotheses by data is presumably a matter of degree.
10. I will not discuss here the various methods that systematics use to test such assumptions about character polarity, on which see Sober (1988c). Also a fine point that will not affect my conclusions is worth mentioning. The two evolutionary assumptions just mentioned entail, not just that *a* common ancestor of the three taxa lacked wings, but that this was the character state of the three taxa's *most recent common ancestor.* This added assumption is useful for expository purposes, but is dispensable. I do without it in Sober (1988b).
11. The complements of e_i and r_i do not represent the probabilities of stasis. They represent the probability that a branch will end in the same state in which it began. This could be achieved by any even number of flip-flops.
12. I will note, however, that assigning e_i and r_i the same value for all the i branches is implausible, if the probability of change in a branch is a function of the branch's temporal duration. In addition, there is a simplification in the treatment of likelihood presented here that I should note. The $(AB)C$ hypothesis represents a family of trees, each with its own set of

branch durations. This means that the likelihood of the $(AB)C$ hypothesis, given a model of character evolution, is an *average* over all the specific realizations that possess that technology. See Sober (1988b) for details.

13. Of course, by packing the auxiliary assumptions into the hypothesis under test, one can obtain a new case in which the 'hypotheses' have well-defined likelihoods without the need to specify still further auxiliary assumptions. My view is that this logical trick obscures the quite separate status enjoyed by an assumed model and the hypotheses under test.
14. The principal claim of this paper is not that 'parsimony' is ambiguous. I see no reason to say that the term is like 'bank'. Rather, my locality thesis concerns the *justification* for taking parsimony to be a sign of truth.
15. I have tried to develop this thesis about evidence in Sober (1988a) and in Essay 8.

REFERENCES

Carnap, R. (1950) *Logical Foundations of Probability* (University of Chicago Press).

Dawkins, R. 1976. *The Selfish Gene* (Oxford University Press).

Edwards, A., and Cavalli-Sforza, L. 1963. 'The Reconstruction of Evolution', *Ann. Human Genetics, 27*, 105.

Edwards, A., and Cavalli-Sforza, L. 1964. 'Reconstruction of Evolutionary Trees' in V. Heywood and J. McNeil (eds.), *Phenetic and Phylogenetic Classification* (New York Systematics Association Publications), *6*, 67–76.

Felsenstein, J. 1983. 'Parsimony in Systematics', *Annual Review of Ecology and Systematics, 14,* 313–33.

Fisher, R. 1930. *The Genetical Theory of Natural Selection* (Reprinted; New York: Dover, 1958).

Haldane, J. 1932. *The Causes of Evolution* (Reprinted New York: Cornell University Press, 1966).

Hempel, C. 1965. *Aspects of Scientific Explanation and Other Essays in the Philosophy of Science* (New York: Free Press).

Hennig, W. 1966. *Phylogenetic Systematics* (Urbana: University of Illinois Press).

Jaynes, E. 1968. 'Prior Probabilities', *IEEE Trans. Systems Sci. Cybernetics, 4,* 227–41.

Jeffreys, H. 1957. *Scientific Inference* (Cambridge University Press).

Miller, R. 1987. *Fact and Method* (Princeton University Press).

Popper, K. 1959. *The Logic of Scientific Discovery* (London: Hutchinson).

Quine, W. 1966. 'On Simple Theories of a Complex World', in *Ways of Paradox and Other Essays* (New York: Random House).

Rosenkrantz, R. 1977. *Inference, Method and Decision* (Dordrecht: D. Reidel).

Rosenkrantz, R. 1979. 'Bayesian Theory Appraisal: A Reply to Seidenfeld', *Theory and Decision, 11,* 441–51.

Salmon, W. 1984. *Scientific Explanation and the Causal Structure of the World* (Princeton University Press).
Seidenfeld, T. 1979. 'Why I am not a Bayesian: Reflections Prompted by Rosenkrantz', *Theory and Decision, 11,* 413–40.
Sober, E. 1975. *Simplicity* (Oxford University Press).
Sober, E. 1981. 'The Principle of Parsimony', *British Journal for the Philosophy of Science, 32,* 145–56.
Sober, E. 1984. *The Nature of Selection* (Cambridge, Mass.: MIT Press).
Sober, E. 1987. 'Explanation and Causation: A Review of Salmon's *Scientific Explanation and the Causal Structure of the World*', *British Journal for the Philosophy of Science, 38,* 243–57.
Sober, E. 1988a. 'Confirmation and Law-Likeness', *Philosophical Review, 97,* 617–26.
Sober, E. 1988b. *Reconstructing the Past: Parsimony, Evolution, and Inference* (Cambridge, Mass.: MIT Press).
Sober, E. 1988c. 'What is Evolutionary Altruism?', *Canadian Journal of Philosophy Supplementary Volume, 14,* 75–99.
Van Fraassen, B. 1980. *The Scientific Image* (Oxford University Press).
Wade, M. 1978. 'A Critical Review of Models of Group Selection', *Quarterly Review of Biology, 53,* 101–14.
Williams, G. C. 1966. *Adaptation and Natural Selection* (Princeton University Press).
Wilson, D. 1980. *The Natural Selection of Populations and Communities* (Menlo Park, CA: Benjamin/Cummings).
Wright, S. 1945. 'Tempo and Mode in Evolution: a Critical Review', *Ecology, 26,* 415–19.

8

The principle of the common cause

In *The Direction of Time,* Hans Reichenbach (1956) stated a principle that he thought helps guide nondeductive inference from an observed correlation to an unobserved cause. He called it *the principle of the common cause.* Wesley Salmon (1975, 1978, 1984) subsequently elaborated and refined this idea.

Reichenbach thought that philosophy had not sufficiently attended to the fact that quantum mechanics had dethroned determinism. His principle was intended to give indeterminism its due; the principle bids one infer the existence of a cause, but the cause is not thought of as part of a sufficient deterministic condition for the effects. It is ironic that as plausible as Reichenbach's principle has seemed when applied to macro-level scientific phenomena and to examples from everyday life, it has been refuted by ideas stemming from quantum mechanics itself. The very idea that Reichenbach wanted to take seriously has come to be seen as his idea's most serious problem.

In the light of this difficulty, Salmon (1984) has modified his endorsement of the principle. It applies, not to all observed correlations, but to correlations not infected with quantum mechanical weirdness. Salmon's fallback position has been that Reichenbach's principle would have been quite satisfactory, if quantum mechanics had not made such a hash of our "classical" picture of causality. This reaction, I think it fair to say, has been the standard one, both for philosophers like Suppes and Zinotti (1976) and van Fraassen (1982) who helped make the case that the principle as originally formulated was too strong and for philosophers like Salmon who had to take this fact into account.

In this paper, I want to describe some consequences of thinking about this principle from an entirely different scientific angle. My interest is to see how the principle fares, not in connection with quantum theory, but with respect to the theory of evolution. Evolutionists attempt to infer the phylogenetic relationships of species from facts about their sameness and difference. We observe that chimps and human beings are similar in a great many respects; we infer from this that there exists a common ancestor of them both, presumably one from whom these shared traits were inherited as homologies. This very approximate description of how phylogenetic inference proceeds suggests that it should be an instance of the problem that Reichenbach and Salmon's principle was meant to address. Inferring common ancestry is a case of postulating a common cause.[1]

In what follows, I'll begin by saying what the principle of the common cause asserts. I'll then argue that the principle has a very serious defect – one that has nothing to do with quantum mechanics, but with a familiar fact about nondeductive inference that discussions of the principle of the common cause have not taken sufficiently to heart.

Let's begin with one of Reichenbach's simple examples. Suppose we follow a theatre company over a period of years and note when the different actors get gastro-intestinal distress. We observe that each actor gets sick about once every hundred days. We follow the company for long enough to think that this frequency is stable – reflecting something regular in their way of life. So by inferring probabilities from frequencies, we say that each actor has a chance of getting sick on a randomly selected day of 0.01. If each actor's illness were independent of the others, then we would expect that two actors would both get sick about once every 10,000 days. But suppose our observations show that if one actor gets sick, the others usually do as well. This means that the probability that two get sick is greater than the product of the probabilities that each gets sick ($1/100$ being greater than $1/100 \times 1/100$). We have here a positive correlation.

The principle of the common cause says that this correlation should be explained by postulating a cause of the two correlates. For example, we might conjecture that the actors always dine together, so that on any given day, all the actors eat tainted food or none of them does. Suppose, for the sake of simplicity, that eating tainted food virtually guarantees gastro-intestinal distress, and also that a person rarely shows the symptom without eating tainted food. Then if we further suppose that the food is tainted about once every hundred days, we will have explained the phenomena: We will have shown why each actor gets sick about once every hundred days and also why the actors' illnesses are so highly correlated.

The postulated common cause has a special probabilistic property; it renders the observed correlates conditionally probabilistically independent. The common cause is said to *screen off* one effect from the other. We can see what this means intuitively by considering the idea of predicting one effect from the other. Given that the two actors' states of health are positively correlated, the sickness or wellness of one on a given day is a very good predictor of the sickness or wellness of the other. However, if we know whether the shared food is tainted or not, this also helps predict whether a given actor is sick or well. The idea that the cause screens off one effect from the other can be stated like this: If you know whether the food is tainted or not, additional knowledge about one actor provides no further help in predicting the other's situation.

The principle of the common cause, then, has two parts. First, it says that observed correlations should be explained by postulating a common cause. Second, it says that the common cause hypothesis should be formulated so that the conjectured cause screens off one effect from the other.

Notice that I have formulated the principle as a piece of epistemology, not ontology. I did not have it say that each pair of correlated observables have a screening-off common cause. It seems that this existence claim is thrown in doubt by quantum mechanics. But the bearing of this physical result on the epistemological formulation is less direct. Perhaps it is sound advice to postulate a common cause of this sort, even if we occasionally make a mistake. I suspect that may be why Salmon, van Fraassen, and others have not doubted that the principle of the common cause is sensible, once one leaves the quantum domain.

This doesn't mean that the physical result has absolutely no methodological upshot. It has a very modest one. Suppose we observe a correlation and postulate a common cause that satisfies a few simple "classical" requirements (including, but not limited to, Reichenbach's principle). This combination of observed correlation and theoretical postulate implies a further claim about observables – namely, Bell's (1965) inequality. Apparently, when this prediction is tested in the right sort of principle experiment, the inequality comes out false. So by *modus tolens,* we are forced to reject at least one of our starting assumptions. Since the observed correlation is well attested, our search for a culprit naturally gravitates to the theoretical postulates. Reichenbach's principle may have to be discarded; at least it isn't clear that the principle should be retained.[2]

This means that if you face a situation in which you have certain beliefs about the observables, it might *not* be reasonable for you to postulate a screening-off common cause. But this is a very small challenge to the principle construed methodologically, since, by and large, we do not find ourselves in that epistemic circumstance. We are not stopped short, nor should we be, in the theatre troupe example by thoughts of Bell's inequality.

So much for the principle of the common cause. Now I'll sketch a few facts about the problem of phylogenetic inference. This problem does not take the form of deciding whether two species are related. This is usually assumed at the outset. Rather, the real question concerns propinquity of descent. We want to know, for example, whether human beings are more closely related to chimps than they are to gorillas; that is to say, whether human beings and chimps have an ancestor that is not also an ancestor of gorillas.

It is now very controversial in evolutionary theory how such questions are to be decided. There are several methods in the field; these sometimes yield contradictory results. So a heated debate has ensued, with biologists trying to justify their pet methods and to puncture the pretensions of rival approaches. This controversy has mainly focused on the issue of what the various methods assume about the evolutionary process. The idea that correlation of characters is evidence for closeness of evolutionary relationship is anything but widely accepted; it has its partisans, but it certainly does not represent a consensus, there presently being no such thing.[3]

So without emersing ourselves too deeply in the details of this dispute, we can nevertheless discern a simple surface feature that is not without its significance. Biologists of every stripe concede that correlation is evidence for propinquity of descent only if certain contingent assumptions are made about the evolutionary process. There is dispute about what these are and about whether this or that assumption is plausible. But it goes without saying in the biology that correlation is not a self-evident indicator of common ancestry, but needs to be defended by a substantive argument.

This may, of course, be because the biologists are misguided. Perhaps the principle does have some sort of *a priori* claim on our attention, quantum difficulties notwithstanding. But I think that this is a mistake. My main claim against the principle will be that correlations are evidence for a common cause only in the context of a background theory.

We do not need recondite evolutionary ideas to grasp this point. There are many correlations that we recognize in everyday life that do not cry out for common cause explanation. Consider the fact that the sea level in Venice and the cost of bread in Britain have both been on the rise in the past two centuries. Both, let us suppose, have monotonically increased. Imagine that we put this data in the form of a chronological list; for each date, we list the Venetian sea level and the going price of British bread. Because both quantities have increased steadily with time, it is true that higher than average sea levels tend to be associated with higher than average bread prices. The two quantities are very strongly positively correlated.

I take it that we do not feel driven to explain this correlation by postulating a common cause. Rather, we regard Venetian sea levels and British bread prices as both increasing for somewhat isolated endogenous reasons. Local conditions in Venice have increased the sea level and rather different local conditions in Britain have driven up the cost of bread. Here, postulating a common cause is simply not very plausible, given the rest of what we believe (Sober 1987).

Let me generalize this point by way of an analogy. Hempel (1945) formulated the raven paradox by asking why observing black ravens confirms the generalization that all ravens are black. He asked that we address this problem by adopting a methodological fiction – we are to imagine ourselves deprived of all empirical background information. In this barren context, we then are to show why observing a positive instance confirms a generalization.

Philosophers like Good (1967), Suppes (1966), Chihara (1981), and Rosenkrantz (1977) responded by showing that there are cases in which background knowledge can make it the case that positive instances *dis*-confirm a generalization. Hempel (1967) replied by saying that these examples were not to the point, since they violated the methodological fiction he had proposed. Good (1968) answered by claiming that if we have no background information at all, then positive instances have no evidential meaning. There is no saying whether they confirm, disconfirm, or are neutral, unless empirical background assumptions are set forth.

The principle of the common cause has an important point of similarity with the idea that positive instances always confirm generalizations. Both, I suggest, are false because they are overstated. Correlations always support a common cause hypothesis no more than positive instances always confirm a generalization.

A fallback position, analogous to the one Hempel adopted by proposing his methodological fiction, suggests itself. Rather than saying that it is *always* plausible to postulate a common cause, why not propose that it is plausible to do so, *if* one knows nothing at all about the subject matter at hand? My view is that there is no assessment of plausibility, except in the light of background assumptions. If there is anything that deserves to be called a principle of the common cause, it must be stated as a conditional; we must say *when* a common cause explanation is preferable to an explanation in terms of separate causes.

I just described the correlation between Venetian sea levels and British bread prices as better explained by postulating separate causes. However, it might be objected that the two correlates do have a common cause, which we can recognize if we are prepared to talk vaguely enough and to go far enough back in time. For example, perhaps the emergence of European capitalism is a common cause. This brings me to a second issue concerning the Reichenbach / Salmon principle, one which the biological problem helps clarify.

I claimed before that a problem of phylogenetic inference is never formulated by focusing on two species and asking if they are related. Rather, the systematist's problem is always a comparative one. We'd like to know, in our running example, whether human beings and chimps are more closely related to each other than either is to gorillas.

I suggest that a parallel point applies to causality in general. Suppose that the causal relation between token events is transitive. Suppose further that the universe we inhabit stems from a single origination event. If we hold that every event in our universe, save this first Big Bang, has a cause, then we have here a general structure of which the phylogenetic situation is a special case. It now is trivially true that every pair of events has a common cause.

In this hypothetical circumstance, would it suddenly become meaningless to consider the credentials of a "separate" cause explanation? I think not. But this means that a nontrivial question about common causes must have more content than merely inquiring whether two events have a common cause.

Returning to Venice and Britain, perhaps our question should be stated as follows: Does the increase in Venetian sea levels and in British bread prices have a common cause that is not also a cause of French industrialization? Perhaps the answer here is *no.* But if we ask, instead, does the increase in Venetian sea levels and in British bread prices have a common cause that is not also a cause of Samoan migrations, the answer may be *yes.* Varying the third term in the question about common causes changes the problem. By analogy with the phylogenetic case, we can see that this term plays the role of specifying the level of resolution at which the problem about common causes is posed. When I said before that the Venice / Britain correlation is not plausibly explained by postulating a common cause, I had in mind a particular level of analysis, but there are others.

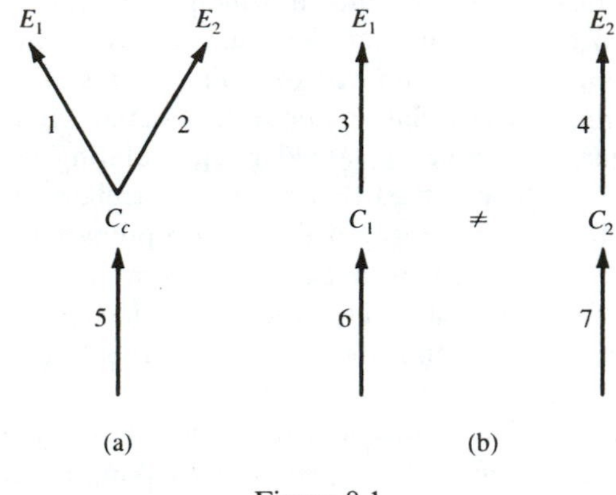

Figure 8.1

Earlier, I segmented the principle of the common cause into two parts. First, there is the idea that observed correlations should be explained by postulating a common cause. Second, there is the demand that the common cause be characterized as screening off each of the correlates from the other. My Venice / Britain example was intended to challenge the first of these ideas. It now is time to see whether a postulated common cause should be thought of as inducing the special probabilistic relationship of *screening off.*

I'll begin by outlining a circumstance in which the screening-off idea makes perfect sense. Let's imagine that we are considering whether the observed correlation between E_1 and E_2 in Figure 8.1 should be explained by postulating a common cause or a separate cause. These are event types, whose frequencies, both singly and in combination, we wish to explain. Figure 8.1a describes the common cause and 8.1b the separate cause pattern of explanation.

How are we to evaluate the quality of explanation provided? The idea of *likelihood* is intuitively appealing, at least as a start. In the theatre troupe example, it is natural to reason as follows: If the actors share their meals, then it is not very surprising that their sick days are strongly correlated. But if they ate quite separately, then it would be very surprising indeed that their sick days covary so closely. In reasoning in this way, we consider how probable each hypothesis says the observations are. The common cause hypothesis makes the correlation quite probable, whereas the separate cause explanation makes it almost miraculous. We prefer the former explanation because it strains our credulity less.

It is essential to see that in considering how probable the hypotheses make the observations, we are not considering how probable the observations make the hypotheses. Nowhere in the thought just described did we consider how probable it is that the actors get sick together, on the supposition that they eat together. The likelihood of a hypothesis is the probability it confers on the observations; at all costs, it must not be confused with the probability the observations confer on the hypothesis.

So in comparing the patterns of explanation displayed in Figure 8.1, we want to determine what probability the two hypotheses confer on the observations. In the theatre troupe example, we observe that the actors get sick about once every hundred days, but that when one actor gets sick, the other almost always does so as well. Our question, then, is how probable do the explanations represented in Figure 8.1 say these observations are?

There is no answer to this question, because the two hypotheses are incompletely specified. Each branch in the figure can be thought of as a causal pathway. To calculate the probability of the events at the tips, we must know what probabilities to associate with the branches. These are not stated, so no likelihood comparison can be made.

Let us suppose that our assignment of values to these branches is wholly unconstrained by any prior theoretical commitments. We can simply *make up* values and flesh out the common cause and the separate cause explanations in any way we please.

We will find the *best case scenarios* for the common cause and the separate cause patterns. We will assign probabilities to branches, 1, 2, and 5 so as to make the common cause maximally likely. And we will similarly assign probabilities to branches 3, 4, 6, and 7 so as to make the separate cause pattern maximally likely. We then will compare these two best cases. In saying that we choose values so as to maximize the likelihood, we are saying that the hypothesis is set up so as to make the observations as probable as possible.

If we assume that the two separate causes C_1 and C_2 are independent of each other and that all the assigned probabilities must be between 0 and 1 noninclusive, we can obtain a very pleasing result.

We can set up the common cause explanation so that it is more likely than any separate cause explanation. If E_1 and E_2 are correlated, we can assign probabilities in Figure 8.1a so that the implied probabilities of the two singleton events and the implied probability of their conjunction perfectly match their observed frequencies. This is the way to maximize the likelihood. If E_1 occurs about once in a hundred times, then the most likely hypothesis will assign that event a probability of 0.01. If E_1 and E_2 almost always cooccur, so that the conjoint event occurs approximately once every 120 times, then the most likely hypothesis will assign that conjoint event a probability of 1/120. However, this matching of postulated probabilities to sample frequencies cannot be achieved in Figure 8.1b. If we match the singleton probabilities to the singleton frequencies, we will fail to have the conjoint probability match the conjoint frequency. So the best case scenario for the common cause explanation is more likely than the best case scenario for the separate cause explanation.

This kind of reasoning, I think, underlies much of the plausibility of the examples discussed by Reichenbach and Salmon. We *invent* hypotheses of common and separate cause and then compare them. There is nothing especially implausible about the idea that actors should always eat together, or that they should eat separately, so the common cause and the separate cause explanations are not radically different on that count. That is, the *probability* of common meals or separate meals is not a consideration of likelihood. Reichenbach and Salmon were, I think, picking up on the fact that in this special context, the common cause explanation has a virtue that the separate cause explanation cannot claim. That virtue is pinpointed by the concept of likelihood.

However, this is a special case, for at least two reasons. First, we have already seen, in the Venice/Britain example, that there is more to be considered than likelihood. Even though the supposition of a common cause might explain the correlation of sea levels and bread prices, it is immensely implausible to suppose that there is such a common cause. This judgment seems to go beyond the dictates of likelihood. To bring this point to bear on the acting example, merely suppose that we are told that it is enormously improbable that the actors took many of their meals together. Of course, if they had done so, that might be a beautiful explanation of the correlation. But we might nevertheless have to reject this likely story because it is immensely improbable.

The second limitation arises even if we restrict ourselves to the issue of likelihood. The problem is that we filled in the branch probabilities in Figure 8.1 in any way we pleased. However, suppose we have background information that constrains how these can be specified. Suppose I tell you that when these actors eat together, the food is checked beforehand, so that the probability of tainted food getting through to them is one in a billion. This information makes the common cause explanation plunge in likelihood. Even if it were true that the actors take their meals together, it isn't so clear any more that this is a very

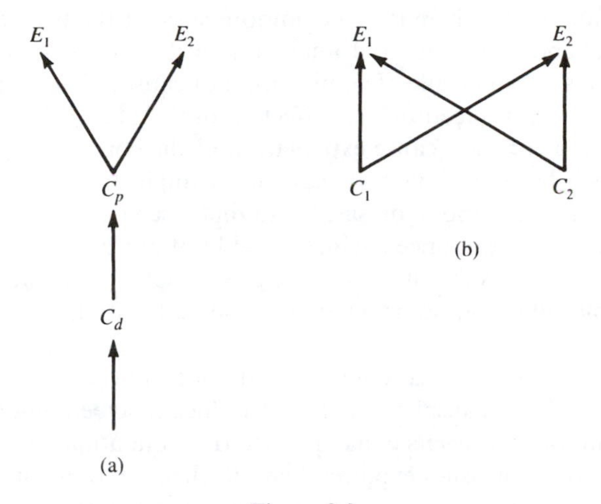

Figure 8.2

plausible explanation of the observations. If the probabilities that figure in the separate cause explanation are fleshed out by other kinds of background information, it may even emerge that the separate cause explanation is more likely after all.

So highly specific background information may prevent the common cause hypothesis from being more likely. I now want to show that there can be quite general theoretical reasons for thinking that plausible assignments of probabilities to elaborate the common cause structure of Figure 8.1a should *not* induce a screening-off relation. Let's consider first the causal set-up depicted in Figure 8.2a. Here we have a distal cause (C_d) producing a proximal cause (C_p), which then causes the two correlated effects (E_1 and E_2). Let's assume, first, that all probabilities are between 0 and 1 noninclusive and that the proximate cause makes a difference in the probability of each effect. If we further assume that the proximal cause screens off the distal cause from the effects and that the proximal cause screens off the effects from each other, it follows that the distal cause does *not* screen off the effects from each other.[4] This means that if I postulate a common cause of the two effects, and I imagine that the cause is distally related to the effects as just described, then I should *not* formulate my common cause hypothesis so that it involves screening off. But this, we can be sure, will not prevent me from testing that hypothesis against other competing hypotheses, and perhaps thinking that it is plausible.

The same lesson arises from the set-up depicted in Figure 8.2b. Suppose I think that if there is one common cause of the two effects, then there probably is another. If I imagine that these causes contribute to their effects by Mill's rule of composition of causes, then neither cause, taken alone, will screen off one effect from the other.[5] So if I propose to develop a common cause explanation of the correlation, one which I hope will be true, though it may be incomplete, I will not want to formulate my description of, say, C_1, so that it screens off the two effects from each other. But once again, this will not prevent me from finding that the hypothesis that fleshes out the idea that C_1 is a common cause is more plausible than some separate cause explanation I choose to consider.

This point does not go counter to the idea that a totally complete picture of *all* the causal facts should induce a screening-off relation. That intuition receives its comeuppance from quantum mechanics, not from the kinds of issues explored here. Rather, the thought behind the examples displayed in Figure 8.2 is that at any stage of inquiry in which we introduce a common cause without the presumption that we are laying out the whole causal story, the demand for screening-off may well be misplaced. For the fact of the matter is that most common causes in the real world do not screen off.

I do not deny that when causal models are developed, it is often natural to formulate them with proximal causes (like C_p in Figure 8.2(a)) screening off effects from each other and with proximal causes screening off distal causes from their effects. But this should be regarded as a useful fiction, not as reflecting some fundamental principle about how a common cause must be understood.

It may be objected that my discussion of the examples depicted in Figure 8.2 does not go counter to the principle of the common cause. After all, that principle says that it is reasonable to think that there exists a screening-off common cause; it does not assert that every common cause screens off. I concede that if the principle did no more than recommend this existence claim, then my complaints would not apply. However, I then would argue that the principle has very little to do with the ongoing process of inventing and evaluating causal explanations. It is worth remembering that, in the work of Reichenbach and Salmon, the principle *does* do more; it tells one how to compare the credentials of fleshed out common cause and separate cause explanations. The point of Figure 8.2 is to show why screening off can count as a weakness rather than a strength in common cause explanation.

So my claim against the principle of the common cause is two-fold. Not only do correlations not always cry out for explanations in terms of a common cause; in addition, common cause explanations are often rendered implausible by formulating them so that they screen off the correlates from each other.

When the principle of the common cause is applied to favorable examples, it seems natural and compelling. When one further sees that it is cast in doubt by quantum mechanics, this merely reinforces the conviction that it reflects some very deep and fundamental fact about how we structure our inferences about the world. However, I have been arguing that both the favorable examples and the physical counterexamples make the principle look more fundamental than it really is. It does not state some ultimate constraint on how we construct causal explanations. Rather, it has a derivative and highly conditional validity.

It is useful to represent the problem of comparing the credentials of a common cause (CC) and a separate cause (SC) hypothesis as a variety of Bayesian inference. This does not mean that Bayesianism is without its problems. But in a wide range of cases, a Bayesian representation describes in an illuminating way what is relevant to this task of hypothesis evaluation. Bayes' theorem allows us to calculate the posterior probabilities of the two hypotheses, relative to the evidence (E) they are asked to explain, as follows:

$$\Pr(CC/E) = \Pr(E/CC)\Pr(CC)/\Pr(E)$$
$$\Pr(SC/E) = \Pr(E/SC)\Pr(SC)/\Pr(E).$$

Since the denominators of these two expressions are the same, we find that the common cause explanation is more probable than the separate cause explanation precisely when

$$\Pr(E/CC)\Pr(CC) > \Pr(E/SC)\Pr(SC).$$

We might imagine that the evidence E considered here consists in the observed correlation of two or more event types.

When will a common cause explanation be preferable? I doubt that one can say much about this question in general, save for the inequality stated above. This is as much of a general principle of the common cause as can be had. If the likelihoods are about the same, then a difference in priors will carry the day; and if the priors differ only modestly, a substantial difference in likelihood will be decisive.

An example of the first kind is afforded by our Venice/Britain example: a common cause explanation of the correlation of sea levels and bread prices can be concocted, one which makes the observed correlation as probable as any separate cause explanation could achieve. This explanation may therefore have a high likelihood, but it may be implausible on other grounds. A Bayesian will represent this other sort of defect by assigning the hypothesis a low prior probability.[6]

An example of the second kind is provided by Reichenbach's example of the theatre troupe. Our background information may tell us that there is no substantial difference in prior probability between the hypothesis that the actors usually take their meals together and the hypothesis that they usually eat separately. What makes the former hypothesis overall more plausible in the light of the observations is the way it predicts the covariance of sick days. Here, it is likelihood that is decisive in comparing common cause and separate case explanations.

So what I'm suggesting is that the principle of the common cause is not a *first* principle. It is not an ultimate and irreducible component of our methodology of nondeductive inference. What validity it has in the examples that have grown up around this problem can be described pretty well within a Bayesian framework. And of equal importance is the fact that a Bayesian framework allows us to show why the preference for common cause explanations can sometimes be misplaced. It is specific background information about the empirical subject matter at hand, not first principles about postulating causes, that will settle, in a given case, what the relevant facts about priors and likelihoods are.

The idea that some first principle dictates that postulating common causes is preferable to postulating separate causes has had a long history. Indeed, in the *Principia* Newton gives this idea pride of place as the first of his "Rules of Reasoning in Philosophy":

> *We are to admit no more causes of natural things than such as are both true and sufficient to explain their appearances.* To this purpose the philosophers say that Nature does nothing in vain, and more is in vain when less will serve; for Nature is pleased with simplicity and affects not the pomp of superfluous causes.

The principle of the *common* cause (with emphasis on the penultimate word) deserves to be situated in this tradition: It is a parsimony principle intended to constrain causal explanation.

There is no magical primitive postulate that says that one cause is always better than two. Everything depends on the details of what the one is and what the two are. Nor, I think, is there any principle that says that in a theoretically barren context – one in which the investigator knows nothing about the subject under investigation – that one cause is preferable to two. This I take to be a natural application of what we have learned about induction and the raven paradox to this problem concerning inference to the best explanation.

I mentioned earlier that it is controversial, to say the least, to take correlation as the right way to estimate propinquity of descent. Another controversial idea about phylogenetic inference goes by the name of parsimony. It holds that the best phylogenetic hypothesis is the one that requires the fewest multiple originations. When we find that human beings and chimps are similar, we prefer a genealogical hypothesis that permits these similarities to be derived as homologies from a common ancestor, rather than as having originated independently. The intuition that common causes offer better explanations than separate causes again is very much at work here.

The ideas sketched above about the principle of the common cause apply with equal force to the philosopher's general notion of parsimony and to the specific idea of parsimony that biologists have suggested for use in the context of phylogenetic inference. If parsimony is the right way to infer genealogical relationships, this must be because certain contingent process assumptions are true concerning how evolution works. What is now less than evident, I think, is what those contingent assumptions actually are. But that there must be nontrivial assumptions of this sort is the lesson I draw from this analysis of the strengths and limitations of the Reichenbach / Salmon idea. There is a useful corrective to the formulations we have come to call *the principle of the common cause* and *the principle of parsimony;* it is the idea that *only in the context of a background theory do observations have evidential meaning.*

APPENDIX A

Claim: If (i) a proximate cause (C_p) screens off one effect (E_1) from the other (E_2), (ii) the proximate cause screens off a distal cause (C_d) from each effect, (iii) all probabilities are intermediate, and (iv) the proximate cause makes a difference in the probability of each effect, then the distal cause does not screen off the one effect from the other.

We begin with the following probabilities:

$$\begin{aligned} P(C_p / C_d) &= p \\ P(E_1 / C_p) &= a \\ P(E_1 / -C_p) &= b \\ P(E_2 / C_p) &= x \\ P(E_2 / -C_p) &= y \end{aligned}$$

Conditions (i) and (ii) imply that $P(E_1 \,\&\, E_2 / C_d) = P(E_1 / C_d)P(E_2 / C_d)$ precisely when

$$pax + (1 - p)by = [pa + (1 - p)b][px + (1 - p)y].$$

We will disprove this equality by reductio. First, we simplify it to

$$p(1 - p)ax + p(1 - p)by = p(1 - p)ay + p(1 - p)bx.$$

Given assumption (iii), this becomes

$$a(x - y) = b(x - y),$$

whose falsehood is guaranteed by (iv), which I interpret to mean $a \neq b$ and $x \neq y$.

APPENDIX B

Claim: Let C_1 and C_2 each be a common cause of the effects E_1 and E_2. If (i) the total state of both causes (i.e., a specification of the presence or absence of each) screens off each effect from the other, (ii) all probabilities are intermediate, and (iii) the probability of each effect is an increasing function of the number of causes that are present, then neither cause taken alone screens off one effect from the other.

First, let

$$\begin{aligned} P(C_i) &= c_i \\ P(E_i / C_1 \,\&\, C_2) &= w_i \\ P(E_i / C_1 \,\&\, -C_2) &= x_i \\ P(E_i / -C_1 \,\&\, C_2) &= y_i \\ P(E_i / -C_1 \,\&\, -C_2) &= z_i \end{aligned}$$

for $i = 1, 2$. I will prove the claim for the case in which both causes are present; the other three cases follow the same pattern.

Condition (i) implies that

$$P(E_1 \,\&\, E_2 / C_1) = w_1 w_2 c_2 + x_1 x_2 (1 - c_2),$$

whereas

$$\Pr(E_1 / C_1) = w_1 c_2 + x_1 (1 - c_2),$$

and

$$\Pr(E_2 / C_1) = w_2 c_2 + x_2 (1 - c_2).$$

C_1 screens off the effects from each other precisely when

$$w_1 w_2 c_2 (1 - c_2) + x_1 x_2 c_2 (1 - c_2) = w_1 x_2 c_2 (1 - c_2) + x_1 w_2 c_2 (1 - c_2).$$

Again, we proceed by reductio. Condition (ii) allows this equality to be simplified to

$$w_1(w_2 - x_2) = x_1(w_2 - x_2).$$

I interpret condition (iii) to mean that $w_i > x_i$, $y_i > z_i$ $(i = 1, 2)$, which guarantees that this equality is false. Note that a stronger compositional principle (which might be termed a Millian principle of composition of causes), wherein $w_i - x_i = y_i - z_i > 0$ $(i = 1, 2)$, is not necessary, though it does suffice.

NOTES

1. A fuller treatment of the bearing of phylogenetic inference on the principle of the common cause is provided in Sober (1988), chapter 3.
2. I owe this exposition of the logic of Bell's (1965) argument to van Fraassen (1982).
3. For some indication of the diversity of opinion on the proper methodology of phylogenetic inference, see the essays of Duncan and Stuessy (1984).
4. See the proof in Appendix A.
5. See the proof in Appendix B.
6. Critics of Bayesianism – for example, Edwards' (1972) espousal of likelihood as a sufficient analytical tool – will look for some other representational devices.

REFERENCES

Bell, J. (1965). On the Einstein Podolsky Rosen Paradox. *Physics 1:*196–200.

Chihara, C. (1981). Quine and the Confirmational Paradoxes. In P. French, H. Wettstein, and T. Uehling (eds.), *Midwest Studies in Philosophy 6.* Minneapolis: University of Minnesota Press. 425–52.

Duncan, T., and Stuessy, T. (1984). *Cladistics: Perspectives on the Reconstruction of Evolutionary History.* New York: Columbia University Press.

Edwards, A. (1972). *Likelihood.* Cambridge: Cambridge University Press.

Good, I. J. (1967). The White Shoe is a Red Herring. *British Journal for the Philosophy of Science 17:*322.

Good, I. J. (1968). The White Shoe qua Herring is Pink. *British Journal for the Philosophy of Science 19:*156–57.

Hempel, C. (1945). Studies in the Logic of Confirmation. In *Aspects of Scientific Explanation and Other Essays.* New York: Free Press. 1965.

Hempel, C. (1967). The White Shoe: No Red Herring. *British Journal for the Philosophy of Science 18:*239–40.

Reichenbach, H. (1956). *The Direction of Time.* University of California Press.

Rosenkrantz, R. (1977). *Inference, Method, and Decision.* Dordrecht: D. Reidel.

Salmon, W. (1975). Theoretical Explanation. In S. Korner (ed.), *Explanation.* Oxford: Blackwell. 118–45.

Salmon, W. (1978). Why ask "Why?" *Proceedings and Addresses of the American Philosophical Association 51:*683–705.

Salmon, W. (1984). *Scientific Explanation and the Causal Structure of the World.* Princeton: Princeton University Press.

Sober, E. (1987). Explanation and Causation: A Review of Wesley Salmon's *Scientific Explanation and the Causal Structure of the World. British Journal for the Philosophy of Science 38:*243–57.

Sober, E. (1988). *Reconstructing the Past: Parsimony, Evolution, and Inference.* Cambridge: Bradford / MIT Press.

Suppes, P. (1966). A Bayesian Approach to the Paradoxes of Confirmation. In J. Hintikka and P. Suppes (eds.), *Aspects of Inductive Logic.* Amsterdam: North-Holland Publishing Co. 197–207.

Suppes, P., and Zinotti, M. (1976). On the Determinism of Hidden Variable Theories with Strict Correlation and Conditional Statistical Independence of Observables. In P. Suppes, *Logic and Probability in Quantum Mechanics.* Dordrecht: D. Reidel.

van Fraassen, B. (1982). The Charybdis of Realism: Epistemological Implications of Bell's Inequality. *Synthese 52:*25–38.

9

Explanatory presupposition

Requests for explanation advanced in the form of why-questions contain presuppositions. Most obviously,

(1) Why is it the case that *P?*

presupposes that the proposition *P* is true. In addition, when it is recognised that explanation is a contrastive activity (Drestske 1973, van Fraassen 1980, Garfinkel 1981), an additional source of explanatory presupposition becomes evident. Suppose we seek to answer (1) by showing why *P, rather than some contrasting alternative C,* is true. The resulting question

(2) Why is it the case that *P* rather than *C?*[1]

presupposes that *P and not-C* are both true.

Why-questions with the structure given in (2) make explicit an often tacit relational element in the activity of explaining a proposition. A given proposition may be embedded in different sets of contrasting alternatives, thereby giving rise to different explanatory problems. We may, to use Garfinkel's (1981) example, provide different readings of the question 'why did Willi Sutton rob banks?' One device for bringing out these alternative formulations is in the use of emphasis (either via spoken stress or written italics/underlining). We may wish to explain why Sutton *robbed* banks, why he robbed *banks,* or why *he* (Sutton) robbed banks. These alternatives may be fleshed out in the form of three different questions conforming to pattern (2):

(3a) Why did Sutton rob banks rather than deposit money into them?

I am grateful to Ellery Eells for comments on an earlier draft.

(3b) Why did Sutton rob banks rather than candy stores?
(3c) Why did Sutton, rather than one of his accomplices, rob banks?

Besides the use of emphasis and formulations conforming to pattern (2), there is another way of posing explanatory questions in which the presuppositions are made evident. One may ask 'given *X*, why is it the case that *P*?' The above three questions concerning Sutton may then be expressed as follows:

(4a) Given that Sutton interacted with banks, why did he rob them?
(4b) Given that Sutton was going to rob something, why did he do this to banks?
(4c) Given that someone in Sutton's gang was going to rob banks, why did Sutton do it?

I cite these three linguistic devices – emphasis, the rather-than locution, and the given locution – in order of increasing expressive power. Emphasis allows one to vary only one constituent of the focal proposition at a time. If the focal proposition is that Sutton robs banks, one may emphasize 'Sutton', 'robs', or 'banks', thereby generating the question otherwise formulatable as (3a), (3b), and (3c). But the focal proposition can be embedded in many other contrast spaces. For example, suppose Sutton's gang deliberated over whether Sutton would rob banks or Jones would burn barns, these both being promising sources of cash. One then might ask:

(5) Why was Sutton robbing banks rather than Jones burning barns?

The rather-than locution allows this question to be posed, but emphasis does not.

Similarly, one might wish to know how the fact that Sutton had a tormented adolescence contributed to his career in crime. So one might say: Given Sutton's upbringing, why did he become a robber? But this question is not expressible in terms of the rather-than locution alone.

Regardless of which syntactic device one uses in formulating a why-question, it is clear that two types of presupposition are involved. In general, a question presupposes a given proposition if and only if the proposition must be true for the question to have an answer. Thus, question (3a) presupposes that Sutton robbed banks and also that there was such a person as Sutton. The difference I want to note between these two presuppositions is this. The former cannot be inserted into an answer to the question, on pain of having the reply beg the question.

The latter, however, is insertable. As a partial characterisation of this difference, we may say that a question with the structure given in (2) has *P* and *not-C* as *non-insertable presuppositions;* however, the disjunction, *P or C,* and all that it implies are insertable presuppositions. Each of these insertable propositions is, as it were, a common element shared by both *P* and *C.* The question assumes that *P* and *not-C* are true; in addition, it stipulates that the answer may include any proposition implied by both *P* and its contrasting alternative *C.*[2]

Every why-question has its presuppositions. This was already evident when we considered (1) above. But does every why-question have *insertable* presuppositions? Put differently, this is to ask whether every why-question is *directive* – that is, must it contain nontrivial indications of what a proper answer can assert?

If a question of the form given in (2) is such that *P* and *C* are logical negations of each other, the only implications they have in common will be logical truths. But as the questions considered in (3) show, it is hardly uncommon for *P* and *C* to fail to be so related. More usually, it will turn out that a focal proposition and its contrasting alternative are contraries, not contradictories. What is more, they usually will be contraries only relative to certain contingent assumptions. No law of nature asserts that Sutton could not have robbed both banks and candy stores. These are contrary alternatives, as in (3b), only relative to contingent assumptions about Sutton himself.

Not that a focal proposition and its contrast must always have joint implications beyond the logical truths. A quite natural class of why-questions that do not are ones that concern existence. A biologist may ask why there are tigers; here, the focal proposition simply contrasts with its logical negation. The question does not indicate which of the following reformulations is the one to pose and answer:

Given that there were ancestral felines, why are there tigers?
Given that there were mammals, why are there tigers?
Given that there were animals, why are there tigers?

The question posed – why does the world contain tigers rather than not – does not indicate where the answer should begin. Yet an answer must begin with the existence of *something,* as long as it's true that out of nothing, nothing comes. The verbal question does not settle this, but leaves the matter to be decided by the context of inquiry. Here, then, is a class of intelligible why-questions in which the joint implications of the focal proposition and its contrast are logical truths and nothing else.

This would settle the matter of whether why-questions must always be directive, if we could assume that the insertable presuppositions of (2) are limited to the joint implications of *P* and *C.* But I want to suggest that why-questions of the form given in (2) make two additional assumptions, ones not implied by the conjunction of *P* and *not-C.*[3]

The following question, I claim, has a false presupposition and is thereby unanswerable, even though the focal proposition is true and its contrasting alternative is false:

(6) Why is Kodaly a Hungarian rather than a vegetarian?

Kodaly is a Hungarian and that fact is explicable; the same holds for his not being a vegetarian. Why, then, should (6) be unanswerable? I suggest that (6) implies that these two properties of Kodaly's have a common cause. This is an assumption not implied by Kodaly's being a Hungarian and a non-vegetarian.[4]

Further evidence for this claim comes from considering the given-locution. Questions of the form 'given *X,* why is it the case that *P?*' can have false presuppositions, even when *X* and *P* are both true. I list a timely (and controversial) evolutionary question as an example:

Given that males in species *X* produce many small gametes whereas females produce fewer and larger ones, why do females in species *X* perform a disproportionate amount of parental care?

This question presupposes that the difference between male and female gametes *makes a difference* when it comes to explaining parental care. The question assumes not just that the two propositions are true, but that the one is explanatorily relevant to the other. I suggest that the idea of explanatory relevance here means *causal* relevance.

I want to spell out this presupposition of a common cause in a way that may appear slightly redundant, but will turn out not to be. There are two parts to the common cause presupposition. First, questions that have the form given in (2) presuppose that *P* and *not-C* trace back to a common cause. Second, it is assumed that this common cause discriminates between the alternatives *P* and *C,* i.e., makes *P* more probable than *C.* I will call these the *tracing back* and the *discrimination* assumptions, respectively. Together, they form the common cause presupposition of why-questions.

It is, of course, often no *a priori* matter whether this presupposition is satisfied in a particular case. Only on the basis of empirical evidence can one know that Sutton's bank robbing and Jones' abstention from barn burning trace back to a common cause, as (5) demands. Symmetrically, one may imagine that nationality and diet were bound together in Kodaly's life. But on the assumption that they were quite independent issues, (6) will not be a well-formed question.

Can one ever know *a priori* that the common cause presupposition is satisfied? Of course, if it is not *a priori* that every event has a cause, then the assumption that pairs of events have *common* causes can hardly be *a priori.* But let us reformulate the question: Suppose one assumes that every event has a cause and assumes the resources of logic and definitions. Are there why-questions for which this information settles the issue of whether the common cause presupposition is satisfied? This may seem to be true when *P* and *C* are logical contraries. In this case, whatever makes *P* true apparently must prevent *C* from being so. Isn't one on safe ground in asking why a ball is red all over rather than blue all over, as far as the presupposition of the common cause is concerned? Although this may seem entirely straightforward, in fact it is not. If causality were deterministic, the cause of ball's being red (and thereby not-blue) would have to make red more probable than blue. But since determinism isn't decidable *a priori,* one can never know *a priori* that the common cause presupposition of (2) is satisfied, even when *P* and *C* are logical contraries.

To see why, let us abandon the assumption of determinism and consider the relation of probabilistic causality discussed by Good (1961–2), Suppes (1970), Cartwright (1979), Skyrms (1980), Eells and Sober (1983), and Sober (1984). A positive causal factor, roughly, is one that raises the probability of its effect in every causally relevant background context. So, for smoking to be a positive causal factor in the production of heart attacks, what is required is that each individual should have a higher chance of a coronary if he smokes than he would have if he did not.[5] There is no requirement that smokers always (or even ever) have heart attacks.

This probabilistic construal introduces a special wrinkle into the issue of explanatory presupposition. A positive causal factor, though it must raise the probability of its effect, need not make one of its possible effects more probable than another. That is, although *P* and *not-C* trace back to a common cause, it need not be true that this common cause makes *P* more probable than *C.* This is why I listed the tracing back and the discrimination assumptions separately.

Consider an example, which I'll first sketch formally and then flesh out with some physical details. *P* and *C* are exclusive and exhaustive possible causal consequences (the one positive, the other negative) of a factor F. F raises the probability of P and lowers the probability of C as follows:

$$\Pr(P/F) = .5 > \Pr(P/not\text{-}F) = .25.$$
$$\Pr(C/F) = .5 < \Pr(C/not\text{-}F) = .75$$

Suppose an event that has trait F produces effect P. Since the event made P true, and P and C, we are imagining, are incompatible, the event also made *not-C* true. Yet the occurrence of the event did not make P more probable than C. Invoking that event, I claim, does not explain why *P as opposed to C* came true. The explanation of this fact, besides citing a common cause, must cite a common cause that discriminates. This is why I stated the common cause presupposition in two parts.

Now for a physical realisation of this set-up. I will throw a switch either up (F) or down (*not-F*). If I push up, a fair coin will be tossed, so that the chance of heads (P) and the chance of tails (C) are both .5. If I push the switch down, however, a coin biased against heads will be tossed, so that the chance of heads (P) is .25 and the chance of tails (C) is .75.

Pushing the switch up increases the chance of heads and decreases the chance of tails. I push the switch up, the fair coin is tossed, and it happens to land heads. You see the result and ask why the result was heads rather than tails. This question is unanswerable, even though the occurrence of heads and the nonoccurrence of tails trace back to a common cause, namely my pushing up on the switch. The common cause fails to discriminate.

There is an additional complication concerning the probabilistic construal of causality that affects the formulation of the tracing back condition, but not the ultimate conclusion of this argument. An event may trace back to an earlier event that is its cause, even though the properties of that earlier event did not raise the probability of the effect. Thus, for example, the genetic characteristics of a child trace back to the genetic characteristics of its parents, even when the parental traits were *negative* causal factors with respect to the offspring characteristics. Similarly, if I had pushed down on the switch in the above example, and the coin had (improbably enough) landed heads, it would be true that that (token) outcome traced back to my push, even though this *reduced* the chance of getting heads. But the main point still remains: two events P and *not-C* tracing back to a common cause do not have to trace back to one that discriminates.[6]

There is no need to state the tracing back and the discrimination conditions separately when the system considered is deterministic. For if a factor *deterministically* causes *P* to be true (confers on it a probability of unity), and if *P* and *C* cannot, in the circumstances, be true together, then the factor must *deterministically* prevent *C* from being true (conferring on it a probability of 0). With determinism, a common cause of *P* and *not-C* is a cause that discriminates between *P* and *C*. Without determinism, this need not be so.

Thus, even if we assume that every event has a cause and even if we formulate our why-questions so that *P* and *C* are logical contraries, we still cannot know on the basis of this that the common cause presupposition of (2) is true. Determinism would suffice here, but the issue of determinism cannot be decided *a priori.*

I have used the term 'insert*able*' to describe the presupposition of a common cause as well as the joint implications of *P* and *C*. This is a term of permission, not obligation. An explanation may fail to assert one or the other of these. One interesting example of how this is possible concerns a style of explanation that is not, properly speaking, causal. An *equilibrium explanation,* I have argued elsewhere (Sober 1983), does not say what the cause of the *explanandum* event was, but merely shows how that event would have taken place regardless of which of a range of alternative possible causes had occurred. The present point, however, is just that why-questions presuppose the existence of common causes; this may or may not be endorsed explicitly in the reasonable answers those questions receive.

In conclusion, I should say a bit about the relationship of this set of claims about explanatory presupposition to philosophical discussion of the so-called principle of the common cause (see, for example, Reichenbach 1956, Salmon 1975 and 1978, van Fraassen 1980, and Essay 8 of the present volume). That literature, it should be stressed, was interested in identifying the circumstances in which it is reasonable to advance explanations of a certain sort. My points here, in contrast, do not concern this epistemological issue, but rather are intended to show how the questions we put to nature are loaded with assumptions of a certain sort, whether those assumptions are warranted or not.

This is not the occasion to consider when it is reasonable to treat pairs of events as stemming from a common cause. Details aside, it should be quite obvious that this will not always be appropriate. Yet I have suggested that the why-questions we ask presuppose that a common cause structure is forthcoming. We have control over the contrast space in which we choose to embed a focal proposition. We get to choose the values of *P* and *C* that flesh out schema (2). But no matter how we do this, there is an overarching assumption – one given, I would suggest, by the nature of the activity of explanation itself. The existence of pairs of events linked by common causes is necessary for the possibility of explanation.

NOTES

1. In this essay, I will suppose that rather-than questions contrast a focal proposition *P* with a single alternative *C*. This, of course, is not generally true. The normal form is 'Why is it the case that *P* rather than C_1 or C_2, . . . , or C_n?' My remarks about (2), however, carry over to the case of multiple contrasting alternatives.
2. Notice that I have left it open whether (2) has insertable presuppositions beyond the ones just mentioned. I'll argue below that it does.
3. Although Garfinkel (1983) discusses both the presuppositions of why-questions and the presuppositions that individuals bring to bear in their explanatory projects, he asserts (p. 40) that a question's presuppositions are limited to the joint implications of the focal proposition and its contrasting alternatives. Van Fraassen, on the other hand (1980, pp. 144–145), holds that (2) presupposes that *P* is true and *C* is false and that 'at least one of the propositions that bears its relevance relation to its topic and contrast-class, is also true'. What may count as a relevance relation goes unspecified in his account, although *causality* is offered as one example.
4. Thus, the presuppositions of (2) parallel the implications of assertions that employ terms like 'but' and 'instead'. To say that someone is *F but G* (e.g., poor but honest; bloodied but unbowed) is to imply that there is some sort of conflict between the two traits, and this goes beyond the mere claim that the individual has *F and G.*
5. A weaker, 'Pareto version', of this criterion has been advocated by Skyrms (1980), Eells and Sober (1983), and Sober (1984): The causal factor must raise the probability of its effect in at least one background context and must not lower it in any other.
6. The distinction involved here between individual causality (as a relation between token events) and population-level causality (as a relation between properties in populations) is discussed in Good (1961–2), Skyrms (1980), Eells and Sober (1983).

REFERENCES

Cartwright, N. (1979). 'Causal laws and effective strategies', *Noûs* 13, pp. 419–437.

Dretske, F. (1973). 'Contrastive statements', *Philosophical Review 81,* pp. 411–437.

Eells, E., and Sober, E. (1983). 'Probabilistic causality and the question of transitivity', *Philosophy of Science* 50; pp. 35–57.

Garfinkel, A. (1983). *Forms of Explanation: Rethinking the Questions of Social Theory.* New Haven: Yale University Press.

Good, I. J. (1961–2). 'A causal calculus I and II', *British Journal for the Philosophy of Science 11,* pp. 305–318; *12,* pp. 43–51; *13,* p. 88.

Reichenbach, H. (1956). *The Direction of Time.* Berkeley: University of California Press.

Salmon, W. (1975). 'Theoretical explanation'. In S. Korner (ed.), *Explanation.* Oxford: Blackwell.

Salmon, W. (1978). 'Why ask "Why?" An inquiry concerning scientific explanation'. In *Hans Reichenbach: Logical Empiricist.* Dordrecht: Reidel.

Skyrms, B. (1980). *Causal Necessity.* New Haven: Yale University Press.

Sober, E. (1983). 'Equilibrium explanation', *Philosophical Studies* 43; pp. 201–210.

Sober, E. (1984). *The Nature of Selection: Evolutionary Theory in Philosophical Focus.* Cambridge: Bradford / MIT Press.

Sober, E. (1985). 'Two concepts of cause'. In P. Asquith and P. Kitcher (eds.), *PSA 1984,* volume 2. E. Lansing, Michigan: The Philosophy of Science Association, pp. 405–24.

Suppes, P. (1970). *A Probabilistic Theory of Causality.* Amsterdam: North-Holland Publishing Co.

van Fraassen, B. C. (1980). *The Scientific Image.* Oxford: Oxford University Press.

10

Apportioning causal responsibility

Is this particle's acceleration due more to gravity or to electricity? Classical physics regards this question as well-conceived. We may answer by examining physical details of the system before us and using our knowledge of pertinent laws. Newton's law of gravity tells us how mass produces a gravitational force; Coulomb's law of electricity shows how charge generates an electrical force. Each of these *source laws* can be connected with the *consequence law "F = ma"* to determine which force induces the greater component acceleration.[1]

In this Newtonian case, two questions seem interchangeable: What contribution did gravity (or electricity) make to the particle's acceleration? What difference did gravity (or electricity) make in the particle's acceleration? These questions are simultaneously addressed by investigating two counterfactual questions: How much acceleration would there have been, if the gravitational force had acted, but the electrical force had been absent? How much acceleration would there have been, if the electrical force had acted, but the gravitational force had been absent? Classical particles obey John Stuart Mill's (1859) principle of the composition of causes. The result of the two forces is just the sum of what each would have achieved, had it acted alone. We see what each contributed by seeing what difference each made in the magnitude of the effect.

For the accelerating Newtonian particle, apportioning causal responsibility is a local matter – we can assess the contributions of gravity and electricity just by discovering physical facts about the particle and the forces that affect it. In this context, the contribution a cause makes and the difference it makes seem to be one and the same issue. It would be natural to see here an example of some universal principle concern-

I am grateful to Ellery Eells, Brian Skyrms, and Peter Woodruff for discussion.

ing how science apportions causal responsibility: natural, but unwarranted. For careful examination of how causal responsibility is apportioned in another context casts these questions in an entirely new light.

"Is Jane's height due more to her genes or to her environment?" Biologists have been taught to regard this question as meaningless. The proper way to formulate a question concerning nature and nurture, they often say, is at the population level. Junk the question about Jane and replace it with something like the following: "In the population of U.S. adults, how much of the variation in height is explainable by genetic variation, and how much by environmental variation?"

The example of Jane's height shows that the question of how much a cause contributes to the effect and of how much difference it makes in the effect are two questions, not one. The latter is answerable, though not locally, whereas the former is not answerable at all. There is no such thing as *the* way science apportions causal responsibility; rather, we must see how different sciences understand this problem differently, and why they do so. The particle's acceleration (E) is an effect of gravity (C_1) and electricity (C_2), just as Jane's height (E) is an effect of her genes (C_1) and environment (C_2). But this parallelism belies the following differences:

	Newtonian particle (gravity/electricity)	Ontogeny (nature/nurture)
How much did C_1, C_2 contribute to E?	locally answerable	meaningless
How much difference did C_1, C_2 make in E?	locally answerable[2]	answerable, but not locally
	Questions are equivalent	Questions are not equivalent

The Newtonian approach to apportioning causal responsibility, I have noted, is based on a *theory*. So, too, is the biologist's approach to the nature/nurture dispute. Or, more accurately, it is based on a *technique*. Biologists use a statistical method known as the *Analysis of Variance* (ANOVA) to say which of an array of causal factors explains more and which explains less of the variation in the effect property found in a population. First and foremost, the method is used to explain the variation in height found in the population that Jane inhabits. After describing how this procedure works, I shall argue that ANOVA can be brought to bear on Jane herself, not just on the population she inhabits.

The suggestion will be that ANOVA can be used to describe which causal factor at work in the *singleton case* made the largest difference in the effect. So my proposal will connect a *populational phenomenon* (e.g., variation in height among U.S. adults) with a singleton phenomenon (e.g., Jane's height).

After explaining how the analysis of variance applies to the nature/nurture dispute, I shall try to say why apportioning causal responsibility proceeds so differently, depending on whether one analyzes an organism's ontogeny or a particle's acceleration. Why is a particle's acceleration decomposable in a way that Jane's height is not? This will lead back to Mill's principle of the composition of causes, which is much less central to the difference than might first appear. But, before considering how nature and nurture are disentangled, we must clarify the concept of locality itself.

1. WHAT IS LOCALITY?

What does it mean to affirm or deny that causal responsibility can be apportioned *locally?*

The concept of locality requires more clarification than I shall be able to provide here. So, in the end, I shall have to rely on a somewhat intuitive grasp of the question at issue. Nevertheless, a few remarks may help direct the reader's attention to the right issues.

First, I should note that the problem of locality raised by apportioning causal responsibility is quite different from the one usually discussed in physics. The problem of locality in quantum mechanics has to do with what must be true of two token events if one causes the other. Very roughly, the question is whether a physical signal must proceed from one to the other, passing continuously through a series of intervening space-time points.

The question of causal magnitude is quite different from the problem of whether there can be action at a distance. Suppose, just for the sake of argument, that the physicist's question is correctly answered by some suitable thesis of locality.[3] Even so, this would not answer our question about Jane. Granted, there is a continuous path linking her genes and environment, on the one hand, and her subsequent height, on the other. Locality in the physicist's sense is thereby assured, but the question of apportioning causal responsibility has yet to be addressed. We have yet to see what it means to assess the relative contributions of Jane's environment and genes to her height.

A second problem raised by the question of locality is that it must be formulated in such a way that it is nontrivial. Consider the humdrum fact that causes are rarely in themselves sufficient for their effects. The match's being struck caused it to light, even though it would not have ignited if the match had been wet or if there had been no oxygen in the air. This is the trivial sense in which locality fails; whether one event will cause another depends on features of the world external to them both.

In the nature/nurture controversy, we acknowledge this familiar point when we say that genes are not in themselves enough. For Jane to reach a certain height, she must be raised in an appropriate environment. Genes are no good, unless supplemented by numerous meals. Nor is environment in itself sufficient, since there are genetic configurations that will impede Jane's growth, no matter how much milk she drinks. The standard slogan is that development is the result of a gene/environment interaction.

Be this as it may, our problem is not thereby resolved. Neither genes nor environment are themselves sufficient. But when they conspire to produce Jane's height, how are we to assess their relative contributions? In particular, is this matter resolvable by attending to features of Jane's physical development from zygote to adult? Or must we look to features outside of Jane's ontogeny and the genes and environment that produced it?

It seems clear that the problem of locality depends on how one carves up nature into discrete physical systems. If Jane's sequence of environments and genes comprise the system in question, then locality will be refuted if we find that factors external to them play a role in determining which mattered more. If from the first we think of the population in which Jane resides as the relevant unit of inquiry, however, then we may find that locality is verified. The causal facts about Jane may turn out to depend on the population, but the facts about the population may not depend on anything external to it. If so, whether locality is true or false will turn on whether we pose our question about Jane or about her population. As emphasized earlier, the principal question of interest here concerns "singleton" physical systems (Jane, the particle), not populations of such.

The question of locality presupposes that descriptors of an individual can be divided into ones that are "intrinsic" and ones that are not. Although the predicate "x lives in a genetically homogeneous population" is true of individuals, it nonetheless expresses a nonlocal property of them. I do not believe that the distinction needed here can be drawn syntactically; nor should it be, since the conclusions reached should be language independent. Nor do I have an informative semantic criterion to suggest. As in other problem areas (e.g., how the thesis of determinism should be formulated), we must rely on a somewhat intuitive grasp of what it is for a property to be local.

2. THE ANALYSIS OF VARIANCE

Why do biologists think the nature/nurture question is meaningless unless formulated in terms of population variation? To be sure, it *is* silly to think of Jane's six-foot stature as decomposing into two feet due to genes and four feet due to environment, as if the genes built Jane from the navel up, while the environment took care of what lies below. But one silly suggestion does not show that the question of causal magnitude must be understood nonlocally.

To see why biologists think that apportioning causal responsibility requires a populational analysis, we need to consider the basics of the ANOVA technique. So as to avoid the macabre prospect of experimentally manipulating human beings, let us switch from Jane to a corn plant. It has a certain height; we want to know how genes and environment affected this outcome. As its name suggests, the analysis of variance understands this problem at the population level. Instead of focusing exclusively on the singleton case, we consider different possible corn genotypes ($G_1, G_2, \ldots, G_n$) and different possible environmental conditions ($E_1, E_2, \ldots, E_m$) – amount of fertilizer, for example. There are then $n \times m$ possible "treatments." We might divide a field into different plots, the corn plants on each plot receiving one of the treatments (G_i & E_j, $i = 1, 2, \ldots, n$ and $j = 1, 2, \ldots, m$). Each plot of land would contain the same number of plants; within a plot, the plants would have identical genotypes and would receive identical amounts of fertilizer. We then would record the average height within each plot and enter the result into an n by m ANOVA table:

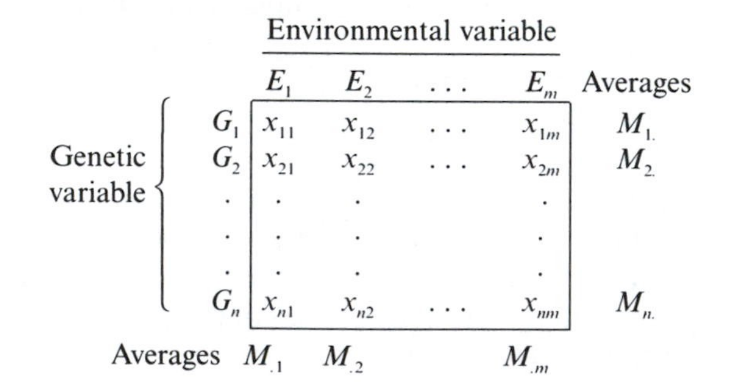

		Environmental variable				
		E_1	E_2	...	E_m	Averages
Genetic variable	G_1	x_{11}	x_{12}	...	x_{1m}	$M_{1.}$
	G_2	x_{21}	x_{22}	...	x_{2m}	$M_{2.}$
	.	.	.		.	
	.	.	.		.	
	.	.	.		.	
	G_n	x_{n1}	x_{n2}	...	x_{nm}	$M_{n.}$
	Averages	$M_{.1}$	$M_{.2}$		$M_{.m}$	

The marginal averages simply record the average height of plants with the same genotype that receive different amounts of fertilizer (the various $M_{i.}$s) and the average height of plants receiving the same amount of fertilizer, but having different genotypes (the various $M_{.j}$s). One final number, not shown in the above table, is the grand mean M, which is simply the average height across all treatments.

It is a matter of arithmetic that the difference between any cell in the above table and the grand mean $(x_{ij} - M)$ must equal the sum of three terms: the so-called *environmental main effect* $(x_{ij} - M_{i.})$, the so-called *genetic main effect* $(x_{ij} - M_{.j})$, and the so-called *gene-environment interaction I* (this last term is a fudge factor, which merely ensures that the sum comes out right):

$$(x_{ij} - M) = (x_{ij} - M_{.j}) + (x_{ij} - M_{i.}) + I.$$

Rearranging the above equation a little, we can show how the average height within any treatment plot must be related to the averages defined on the entire field of plots:

$$x_{ij} = M + (x_{ij} - M_{.j}) + (x_{ij} - M_{i.}) + I.$$

What I have set forth so far is quite unobjectionable.[4] But what use is this partitioning when it comes to the task of causal explanation? I begin with a negative claim: ANOVA does *not* explain the occurrence of singleton effects (i.e., why a given plant has a height of x_{ij}).[5] In the next section, something positive will be defended concerning the relevance of ANOVA to explaining singleton phenomena.

It seems quite clear that the height attained within a given plot does not causally depend on the heights attained in other plots. This is true even though the height within a plot is the sum of terms that represent that plot's deviations from various means (and a corrective factor, I). I would supplement this causal claim with an explanatory one: If you want to know why the corn plants in a given treatment attained the height they did, do not cite the information codified in the analysis of variance formula. The analysis of variance does not identify causes or explain the upshots found in individual cells. (These remarks bear on the "Natural State Model" discussed in Essay 11.)

Consider the fact that a given treatment (G_i & E_j) can be embedded in different experimental designs. For example, I might investigate plants that have the same genotype and place them in different environments. In this case, I shall find that the genetic main effect is zero. As biologists say, all the variance in height is explained by environmental variance. Likewise, I might see how plants of different genotype fare in a single environment. In this case, the environmental main effect will be zero, and all the variance in height will be explained by genetic variance. This shows how misguided it is to interpret the analysis of variance as explaining why a plant in a given treatment cell attained the height it did. If I run the experiment one way, I shall say that the genes played no role; if I run it in the other way, I shall conclude that the environment played no role. And, if I vary both treatments, as I did initially, I shall doubtless conclude that genes and environment both played a role.

An additional reason for rejecting the analysis of variance as a device for identifying the causes that contributed to a singleton effect is provided by considering what this procedure says about characteristics that are universal in a population. Let us switch from corn-plant height to the human characteristic of having a single head. If we consider a population in which everyone has just one head, the analysis of variance will tell us that the genetic and environmental main effects are zero. It would be a mistake to conclude from this, however, that Jane's environment and genes played no role in providing her with a head.

The calculations derived from an ANOVA table do not allow one to deduce what the causes of a given singleton effect are. But this is not to deny that ANOVA can be used to describe the relative contributions that different factors make to explaining the pattern of variation of a trait in a population. Although Jane's genes obviously contribute to her having a head, it may or may not be true that genetic variation helps explain the way that phenotype varies in the population. This is perhaps why biologists have thought of apportioning causal responsibility as fundamentally a population level problem, not one that can be meaningfully addressed for the singleton case. I now want to suggest, however, that the proper use of ANOVA at the population level has relevance to a question about singleton events. One can say which causal factor at work in a singleton event made the greatest difference by embedding that singleton effect in a population, and then judiciously applying the ANOVA to that population.

3. THE COUNTERFACTUAL TEST

I suggest that causal responsibility for Jane's height can be apportioned between her genes and environment by considering two counterfactuals. How tall would Jane be if she had different genes, but the same environment? How tall would Jane be if she had a different environment, but the same genes? Environment is a more powerful influence than genes, if changing her environment would lead Jane's height to depart more from its actual value than would changing her genes.

To answer these two counterfactuals questions, we must elaborate. If Jane were raised in an environment different from the actual one, what would her environment be like? And, if Jane had a different complement of genes, what genes would she in fact have?[6] My suggestion is that the way we fill in the details here is often nonlocal. When this is so, the question of causal magnitude is also a nonlocal one.

When we consider what genes Jane would have had if she had been differently endowed, we do not consider gazelle genes or Martian genes. In some sense, we go to a more "similar possible world." We may consider the other genes that were available from Jane's parents. If these are the same as the ones she actually possesses (i.e., Jane and her parents are all homozygous for the genes in question), we may go back to Jane's grandparents. Or perhaps, rather than tracing the lineage backward in time, we may look at the other alleles present in the local population that Jane inhabits.[7] If there are only two alleles in this local population and Jane has one of them, then it is natural to say that she would have had the other one, if she had had a gene different from the one she actually possesses. If there are several, then we might calculate Jane's expected height as a weighted average of the heights she would have had with each alternative gene, the weighting being supplied by the genes' population frequencies.

Similarly, when we ask what environment Jane would have inhabited had she not inhabited the one she in fact did, we do not imagine her floating weightlessly in outer space, nourished by a diet of candy bars. This is not a "similar possible world." Since human beings inhabit such an incredible variety of environments, it often will be impossible to say which environment Jane would have inhabited, had she pursued a different way of life. If so, a natural strategy is to do what we did above for the case of multiple alleles. We assign probabilities to different possible environments and then calculate Jane's expected height as a weighting over these.

It seems intuitive that, in both these problems, we elaborate our counterfactuals by attending to features of the world which are extrinsic to Jane's actual genes and environment and the height she actually attained. No intrinsic feature of Jane dictates what genes she would have had, if she had had a different genome. The same holds true of her environment.

We can illustrate this idea in a thought experiment akin to Hilary

Putnam's (1975) story about twin Earth. Jane lives in environment E_2 and has genotype G_2. We need to consider what her environment would be, if it differed from E_2, and what her genotype would be, if it differed from G_2.

We shall consider two possible contexts. In the first (I), Jane lives in a population in which the only alternative to E_2 is E_1, and the only alternative to G_2 is G_1. In the second (II), Jane inhabits a population in which the only alternative to E_2 is E_3, and the only alternative to G_2 is G_3. We shall see how this contextual difference leads to contradictory assessments of how we should apportion responsibility for Jane's height between her genes and environment.

The following table describes the height that Jane would have in each of the nine possible gene/environment combinations:

		Environments		
		E_1	E_2	E_3
Genes	G_1	x_{11}	x_{12}	x_{13}
	G_2	x_{21}	x_{22}	x_{23}
	G_3	x_{31}	x_{32}	x_{33}

(Box I encloses x_{11}, x_{12}, x_{21}, x_{22}; box II encloses x_{22}, x_{23}, x_{32}, x_{33}.)

Jane's actual height is given by x_{22}. If she lives in a population of the first type (I), then we assess the relative impact of her genes and environment by comparing x_{12} and x_{21}. If she lives in a population of the second type (II), we must compare x_{23} and x_{32}. Suppose that $x_{22} = x_{12} = x_{23}$. This means that, in situation I, her genes make no difference to her height; if she had had a different genome, she would have had the same height. In situation II, however, it is her environment that makes no difference; if she had had a different environment, she would have had the same height. So, whether genes matter more than environment depends on features of the world external to Jane's own environment and genes. It follows that the question of causal magnitude is nonlocal.

The above table is just what one would obtain by pursuing an analysis of variance on causal factors antecedently identified. One would clone Jane and place her in different environments and place genetically different individuals in the environment she actually inhabits. Or, more precisely, since ethics prohibits this manipulation, one would try to assess what the consequences would be if these things were done. When we make counterfactuals actual, so to speak, we use the analysis of variance to apportion causal responsibility.[8]

4. COMMENSURABILITY AND LOCALITY

In arguing that the analysis of variance correctly apportions causal responsibility, I mean to suggest that it can identify the *difference* that various causes make in the observed effect. This is quite different from assessing how much each contributed. This latter question I interpret as a local one, which the facts of the matter about nature and nurture render unanswerable. Why this is so we can see by inventing a science fiction scenario in which the question of contribution is quite intelligible.

Suppose height were the result of the accumulation of height particles, which organisms could obtain from their environment and also from their genes. Imagine that an individual's height is some increasing function of the number of height particles obtained from all sources. If so, we could look at local facts about Jane and say whether her genes or environment contributed more.

Of course, there are no such things as height particles, but this thought experiment suggests the following (somewhat vague) conjecture: For it to make sense to ask what (or how much) a cause contributes to an effect, the various causes must be commensurable in the way they produce their effects. Height particles would provide this common currency.

The idea of height particles shows the inadequacy of one diagnosis of why locality fails in the nature/nurture case. In the Newtonian example, we can ask what would happen if no gravitational force acted or if no electrical force were at work. In the nature/nurture case, no significant answer is obtained by asking how tall Jane would have been if she had had no genes or no environment. The intelligibility of these questions is not essential, however. If there were height particles, the question of whether genes or environment contributed more would make sense, even though an organism requires genes and an environment of some sort if it is to exist at all.

Nevertheless, this difference between gravity/electricity and nature/nurture is not without its significance. I have claimed that the difference made by gravity and by electricity is a local matter; intrinsic features of the physical system under study determine what would happen if either force had acted alone. But we may embellish the details of this example to show that the issue of locality is, even here, not so straightforward.

We so far have considered a physical system made of a particle and the gravitational (G) and electrical (E) forces that affect its acceleration

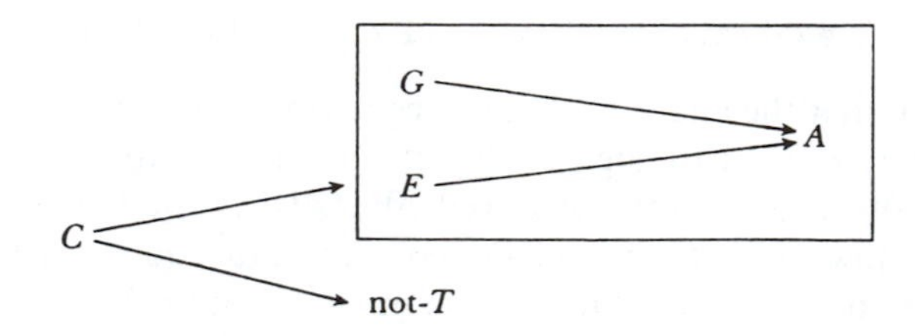

Figure 10.1

(A). Suppose we embed this physical system into a larger context. Let us imagine that some cause C – a switch, say – ensures that a third force T of given magnitude would have come into play if and only if there had been no electrical force (E). The setup is shown in Figure 10.1, with a box drawn around the factors that comprise what I have so far called "the physical system."

There now is an ambiguity in the question of what would have happened if gravity had acted, but electricity had not. We could interpret this, as we have so far, as meaning that gravity would have acted *alone.* Or we could reason that, if electricity had been absent, then the third force (T) would have acted in its stead.

That is, we must decide whether or not we allow counterfactuals to "backtrack."[9] I believe that both interpretations have their place and that it is context that settle which reading is appropriate. If so, there is no univocal answer to the question of whether the difference made by gravity and electricity is locally determined. If we forbid backtracking, then the question of what difference gravity and electricity made reduces to the question of what each would have achieved if it had acted alone. This is a local matter. If we permit backtracking, however, then the question becomes nonlocal. We must consult physical facts extrinsic to the particle and the gravitational and electrical forces acting on it. Thus, the lower-left entry in the table presented in the first section requires that counterfactuals not backtrack.

No similar ambiguity can arise in the nature/nurture case. The reason is that it makes no sense to imagine an individual developing without any environment at all and that an individual will not develop at all, if it has no genes. The ambiguity in the gravity/electricity case was made possible by the fact that we can ask what would have happened if only one of those forces had been present. This provides the nonbacktracking and local reading of the question of what would have happened if one had been present and the other absent.

This local interpretation is not available in the nature/nurture case;

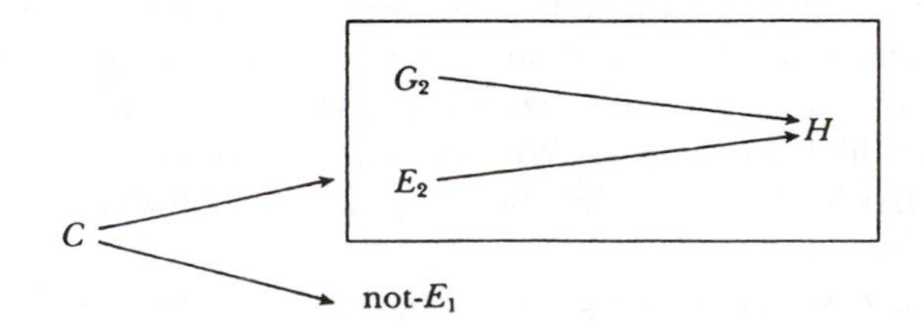

Figure 10.2

we are forced to consider the wider context which dictates what alternative environmental and genetic factors would have been present, if the actual ones had been absent. Here the question of how tall Jane would have been, if she had had a different complement of genes (or a different environment) must be understood in terms of backtracking counterfactuals. Jane's environment E_2 and genes G_2 produce her height (H). Some causal factor (C) presumably determined that she would develop in environment E_2, rather than in E_1. If we reason that Jane would have developed in E_1, if she had not developed in E_2, we are backtracking; we are imagining that the cause that determined that she would grow up in E_2 and not in E_1 would have been different, if she had failed to grow up in E_2. The causal facts may be depicted as in Figure 10.2. I conclude that the question of what difference two causal factors made may be local or nonlocal in the Newtonian case, but that it must be nonlocal when the issue concerns nature/nurture.

Thus, there are at least two differences between the Newtonian particle and the real world of nature and nurture. Causal contribution is well-defined and local in the first, but not in the second. In addition, the two cases part ways over how the difference a cause makes is determined. This can be understood either locally or nonlocally in the Newtonian case, but it must be understood nonlocally in the case of nature and nurture.

It may be thought that the key to the difference between the Newtonian and the biological examples lies in the fact that the former obeys Mill's principle of the composition of causes. As mentioned before, source laws describe which physical properties of the particle system generate this or that force, and the consequence law "$F = ma$" allows the separate contributions of gravity and electricity to be described in a common currency. This generic concept of force is indispensable, if the contributions of each cause are to be comparable. But it now is time to see that it is entirely irrelevant whether the system under study obeys the principle to which Mill called our attention.

Biologists would use the terms "additive" and "nonadditive" to draw Mill's distinction between systems which obey the principle of the composition of causes and ones which are, in his terminology, "heteropathic" (Mill 1859, p. 213). We now may elaborate our story about height particles to show why additivity is neither necessary nor sufficient for locality.

As long as such particles exist, the relative contribution of genes and environment to height will be locally assessable, regardless of the formula that translates number of particles into inches and feet. Suppose that height increases as the number of particles goes from 0 to 50 and declines thereafter. If 40 particles were obtained from the environment and 40 from the genes, we would assign an equal contribution to each source. And this would be correct even though an individual with 80 particles is not twice as tall as an individual with 40.

This shows why additivity is not necessary for locality. To see that it is not sufficient, we must realize that the analysis of variance has no problem describing an additive relation between genes and environment; the data displayed in an ANOVA table are additive, if moving across any row involves adding a certain quantity to one entry to obtain the next and the same holds for moving down any column:

$$x_{i,j+1} = x_{i,j} + c, \quad \text{for some } c \text{ and for all } i,j.$$
$$x_{i+1,j} = x_{i,j} + d, \quad \text{for some } d \text{ and for all } i,j.$$ [10]

The case of two forces (F_1 and F_2) acting together, where their joint effect is simply the sum of what each would have achieved if it had acted alone, is just a special case of this formula:

	not-F_1	F_1
not-F_2	0	c
F_2	d	$c + d$

The entries in a ANOVA table for nature and nurture might be additive in this sense, but this would not give local sense to the question of how much of Jane's height was due to her genes and how much to her environment. I conclude that additivity is neither necessary nor sufficient for locality.

The idea of height particles shows that the question of causal magnitude can be approached both locally and nonlocally with respect to the same physical system. Even if such particles existed, we still could pursue the analysis of variance to obtain a nonlocal answer to the question of causal magnitude. We also could do an intrinsic analysis of the relative contributions of Jane's environment and genes to her height. Indeed, we could obtain very different answers from these two procedures. If her genes produced 50 particles, but her environment only 10, we might judge from this intrinsic perspective that her genes contributed more. If, however, there is little genetic variation in Jane's population, the analysis of variance might conclude that environment made more of a difference than genes.

This shows why saying how much Jane's environment and genes contributed to her height can be a separate matter from saying whether her genes or environment made the greater difference. In the Newtonian case, these two questions are interchangeable and local (provided that counterfactuals are not allowed to backtrack). In the case of nature and nurture, they are different issues. How much difference Jane's environment and genes made to her height is answerable, though not locally. How much each contributed is an unanswerable question, since the empirical fact of the matter (I presume) is that there are no such things as height particles.[11] Apportioning causal responsibility involves two issues, not just one.

In the Newtonian case, the commensurability of gravity and electricity renders the question of causal contribution well-defined and local; in the nature/nurture case, we found neither commensurability nor locality. How much a causal factor contributes to its effect is, I would suggest, an inherently local question. But the question of how much difference a cause makes may be local or not, as is shown by the following example (suggested by Peter Woodruff).

Consider the fact that the distance traveled by a projectile shot from a cannon is influenced by both the muzzle velocity and the angle at which the gun is set. Suppose we fire a cannon and the shot goes half a mile. There is no saying *how much* each factor contributed to this outcome, nor which factor contributed more. The reason is that muzzle velocity and angle setting do not make their contributions in a common currency.

Yet the question of *what difference* the two factors made may be locally or nonlocally assessable, depending on how we elaborate the story. Suppose the physical design of the gun shows that there are just two possible angle settings and two possible powder charges. If so, an intrinsic examination of the system allows us to fill out the relevant counterfactuals. Which factor made the larger difference is now a local matter.

On the other hand, suppose the gun has just one possible angle setting, though it may be packed with different amounts of powder. If the cannon maker designed several guns, each with its own fixed angle of fire, we would have to consider nonlocal facts to say what angle setting the gun would have had, if it had lacked the one it in fact had. Here the fixed setting is analogous to the fact that each of us has just one genetic endowment.

5. CONCLUSION

Causes are not necessary for their effects. It follows that one does not refute a causal claim by showing that the effect would have happened, even if the purported cause had not. Yet when one shifts from the definition of causality to the problem of clarifying the concept of causal magnitude, something like the necessity thesis appears to be correct. If genes and environment are both causal factors influencing Jane's height, then genes have zero magnitude, provided that Jane would have had exactly the same height even if her genes had been different.

I therefore seem to find myself in the paradoxical position of saying that genes can be a cause of height, even if they are judged to have zero magnitude. But perhaps this air of paradox can be dispelled. It is not hard to fathom how causes can fail to be necessary for their effects. Suppose Watson would have shot Moriarty dead if Holmes had not. Holmes' pulling the trigger may have killed Moriarty, even though Holmes' firing was not necessary for Moriarty's death. The point may be put by saying that Holmes' firing, in this case, "made no difference." Causes may make no difference, but they are causes nonetheless.

I also have argued that the relative contribution of a cause and the difference it makes in the effect are separate issues. In the case of our Newtonian particle, this may look like a distinction without a difference. Disentangling the roles of genes and environment, however, shows this distinction to be a real one. Indeed, it is not just the science of ontogeny which draws this distinction. Holmes' pulling the trigger may have made no difference, but he certainly made a contribution to Moriarty's death.

A paradox that is less easy to dispel concerns the question of locality. Waiving as we have the quite separate issues raised by quantum mechanics, we may insist that causality is a local phenomenon. Holmes' firing causes Moriarty's death because there is a continuous process leading from one to the other. The causal relation obtains in virtue of this local circumstance. Watson's standing in the wings is relevant to the question of necessity, but not to the question of causality.

I have argued, however, that, even if causality is local, the magnitude of causality need not be. Jane's own genes and environment locally conspire to produce her adult height. But apportioning causal responsibility is not, in this case, a local matter. This is because it is a factual question, though not a local one, as to whether her genes or environment made the larger difference in her height. And it is not a factual matter at all, much less a local one, as to how much each contributed.

NOTES

1. The distinction between source law and consequence law is elaborated and applied to evolutionary theory in Sober (1984), ch. 1.
2. A proviso will be registered later to defend this entry – namely, that the relevant counterfactuals be nonbacktracking.
3. There are several physical formulations which would have to be sorted out here, as John Earman (1986) explains.
4. For the statistical details of this method, see Sokal and Rohlf (1969). I have omitted mention of an "error" term; this plays a role in ANOVA inference, but does not affect the points about causality to be made in what follows.
5. The following arguments are explored in somewhat more detail in Lewontin (1974) and in Sober (1984).
6. Like all counterfactuals, the two at issue here at times may be vague or indeterminate. To the degree that this is so, I suggest that the question of causal magnitude also is vague or indeterminate.
7. From this point on, I shall imagine that Jane is a haploid organism, so that I can talk of her having a given gene (not two genes) at a locus. This is simply to simplify the exposition.
8. When there is a univocal answer to the question of how tall Jane would be if her genes were different, probabilities need not intrude; when there are multiple alternative possible genomes, a probabilistic weighting is needed. In both these cases, however, there may be disagreement over whether there is a uniquely correct answer to the question of causal magnitude, or whether the "choice of a reference class" is inherently arbitrary. My argument is neutral on this, since, in either case, the thesis of locality fails.
9. I here use the vocabulary, if not the full-blown theory, developed in Lewis (1979).
10. This is a sufficient condition for additivity, not a necessary one; the definition of additivity is just that the interaction term (I) is zero and so the value within each cell is the sum of the grand mean and the two main effects.
11. One could take the position that the questions of how much each factor contributes and of how much difference each makes are equivalent and

nonlocal in the nature/nurture dispute when there are no local determiners of how causal responsibility should be apportioned. The fact that these questions come apart when we imagine a local determiner suggests, however, that the questions are better treated as separate ones from the start.

REFERENCES

Earman, J. (1986). "What is Locality?" In P. Achinstein (ed.), *Theoretical Physics in the Hundred Years Since Kelvin's Baltimore Lectures.* MIT.

Lewis, D. (1979). "Counterfactual Dependence and Time's Arrow." *Noûs 13:* 455–76.

Lewontin, R. (1974). "The Analysis of Variance and the Analysis of Causes." *American Journal of Human Genetics 26:* 400–411.

Mill, J. S. (1859). *A System of Logic, Ratiocinative and Inductive.* Harper.

Putnam, H. (1975). "The Meaning of 'Meaning'." In *Mind, Language, and Reality.* Cambridge University Press.

Sober, E. (1984). *The Nature of Selection: Evolutionary Theory in Philosophical Focus.* MIT Press.

Sokal, R., and Rohlf, J. (1969). *Biometry: The Principles and Practice of Statistics in Biological Research.* Freeman.

11

Evolution, population thinking, and essentialism

1. INTRODUCTION

Philosophers have tended to discuss essentialism as if it were a *global* doctrine – a philosophy which, for some uniform reason, is to be adopted by all the sciences, or by none of them. Popper (1972) has taken a negative global view because he sees essentialism as a major obstacle to scientific rationality. And Quine (1953b, 1960), for a combination of semantical and epistemological reasons, likewise wishes to banish essentialism from the whole of scientific discourse. More recently, however, Putnam (1975) and Kripke (1972) have advocated essentialist doctrines and have claimed that it is the task of each science to investigate the essential properties of its constitutive nature kinds.

In contrast to these global viewpoints is a tradition which sees the theory of evolution as having some special relevance to essentialist doctrines within biology. Hull (1965) and Mayr (1959) are perhaps the two best known exponents of this attitude; they are *local* anti-essentialists. For Mayr, Darwin's hypothesis of evolution by natural selection was not simply a new theory, but a new *kind of theory* – one which discredited essentialist modes of thought within biology and replaced them with what Mayr has called "population thinking." Mayr describes essentialism as holding that

> . . . [t]here are a limited number of fixed, unchangeable "ideas" underlying the observed variability [in nature], with the *eidos* (idea) being the only thing that is fixed and real, while the observed variability has no more reality than the shadows of an object on a cave wall . . . [In con-

Suggestions made by William Coleman, James Crow, Joan Kung, David Hull, Geoffrey Joseph, Steven Kimbrough, Richard Lewontin, Ernst Mayr, Terrence Penner, William Provine, Robert Stauffer, Dennis Stampe, and Victor Hilts helped me considerably in writing this paper.

> trast], the populationist stresses the uniqueness of everything in the organic world. . . . All organisms and organic phenomena are composed of unique features and can be described collectively only in statistical terms. Individuals, or any kind of organic entities, form populations of which we can determine the arithmetic mean and the statistics of variation. Averages are merely statistical abstractions, only the individuals of which the population are composed have reality. The ultimate conclusions of the population thinker and of the typologist are precisely the opposite. For the typologist the type (*eidos*) is real and the variation an illusion, while for the populationist, the type (average) is an abstraction and only the variation is real. No two ways of looking at nature could be more different. (Mayr 1959, pp. 28–9)

A contemporary biologist reading this might well conclude that essentialists had no scientifically respectable way of understanding the existence of variation in nature. In the absence of this, typologists managed to ignore the fact of variability by inventing some altogether mysterious and unverifiable subject matter for themselves. The notion of *types* and the kind of anti-empiricism that seems to accompany it, appear to bear only the most distant connection with modern conceptions of evidence and argument. But this reaction raises a question about the precise relation of evolution to essentialism. How could the *specifics* of a particular scientific theory have mattered much here, since the main obstacle presented by essentialist thinking was just to get people to be scientific about nature by paying attention to the evidence? The problem was to bring people down to earth by rubbing their noses in the diversity of nature. Viewed in this way, Mayr's position does not look much like a form of *local* anti-essentialism.

Other perplexities arise when a contemporary biologist tries to understand Mayr's idea of population thinking as applying to his or her own activity. If "only the individuals of which the population are composed have reality," it would appear that much of population biology has its head in the clouds. The Lotke–Volterra equations, for example, describe the interactions of predator and prey *populations.* Presumably, population thinking, properly so called, must allow that there is something real over and above individual organisms. Population thinking countenances organisms and populations; typological thinking grants that both organisms and types exist. Neither embodies a resolute and ontologically austere focus on individual organisms alone.

Another issue that arises from Mayr's conception of typological and population thinking is that of how we are to understand his distinction between "reality" and "abstraction." One natural way of taking this distinction is simply to understand reality as meaning existence. But presumably no population thinker will deny that there are such things as averages. If there are groups of individuals, then there are numerous properties that those groups possess. The *average* fecundity within a population is no more a property which we invent by "mere abstraction" than is the fecundity of individual organisms. Individual and group properties are equally "out there" to be discovered. And similarly, it is unclear how one could suggest that typologists held that variability is unreal; surely the historical record shows that typologists realized that differences among individuals *exist.* How, then, are we to understand the difference between essentialism and population thinking in terms of what each holds to be "real" about biological reality?

Answering these questions about the difference between essentialist and population modes of thought will be the main purpose of this essay. How did essentialists propose to account for variability in nature? How did evolutionary theory undermine the explanatory strategy that they pursued? In what way does Darwinian biology embody a novel conception of variability? How has population thinking transformed our conception of what is *real?* The form of local anti-essentialism which I will propound in what follows will be congenial to many of Mayr's views. In one sense, then, our task will be to explicate and explain Mayr's insight that the shift from essentialist to populationist modes of thinking constituted a shift in the concept of biological reality. However, I will try to show why essentialism was a manifestly *scientific* working hypothesis. Typologists did not close their eyes to variation but rather tried to explain it in a particular way. And the failure of their explanatory strategy depends on details of evolutionary theory in ways which have not been much recognized.

My approach to these questions will be somewhat historical. Essentialism about species is today a dead issue, not because there is no conceivable way to defend it, but because the way in which it was defended by biologists was thoroughly discredited. At first glance, rejecting a metaphysics or a scientific research program because one of its formulations is mistaken may appear to be fallacious. But more careful attention vindicates this pattern of evaluation. It is pie-in-the-sky metaphysics and science to hold on to some guiding principle simply because *it is conceivable* that there might be some substantive formulation and development of it. Thus, Newtonianism, guided by the maxim that physical phenomena can be accounted for in terms of matter in motion, would have been rejected were it not for the success of particular Newtonian explanations. One evaluates regulative principles by the way in which they regulate the actual theories of scientists. At the same time, I will try in what follows to identify precisely what it is in essentialism and in evolutionary theory that makes the former a victim of the latter. It is an open question to what degree the source of this incompatibility struck working biologists as central. As I will argue at the end of this section, one diagnosis of the situation which seems to have been historically important is much less decisive than has been supposed.

The essentialist's method of explaining variability, I will argue, was coherently formulated in Aristotle, and was applied by Aristotle in both his biology and his physics. Seventeenth- and eighteenth-century biologists, whether they argued for evolution or against it, made use of Aristotle's Natural State Model. And to this day, the model has not been refuted in mechanics. Within contemporary biology, however, the model met with less success. Twentieth-century population genetics shows that the model cannot be applied in the way that the essentialist requires. But the Natural State Model is not wholly without a home in contemporary biology; in fact, the way in which it finds an application there highlights some salient facts about what population thinking amounts to.

An essentialist view of a given species is committed to there being some property which all and only the members of that species possess. Since there are almost certainly only finitely many individuals in any given species,[1] we are quite safe in assuming there is some finitely statable condition which all and only the members of the species satisfy. This could trivially be a list of the spatio-temporal locations of the organisms involved. But the fact that such a condition exists is hardly enough to vindicate essentialism. The essentialist thinks that there is a diagnostic property that any *possible* organism must have if it is to be a member of the species. It cannot be the case that the property in question is possessed by all organisms belonging to *Homo sapiens,* even though there might exist a member of *Homo sapiens* who lacked the trait. It must be necessarily true, and not just accidental, that all and only the organisms in *Homo sapiens* have the characteristic.

However, even this requirement of essentialism is trivially satisfiable. Is it not necessarily true that to be a member of *Homo sapiens* an organism must be a member of *Homo sapiens*? This is guaranteed if logical truths are necessary. But essentialism about biology is hardly vindicated by the existence of logical truths. In a similar vein, if it is impossible for perpetual motion machines to exist, then it is necessarily true that something belongs to *Homo sapiens* if and only if it belongs to *Homo sapiens* or is a perpetual motion machine. This necessary truth is not a truth of logic; it is a result of the theory of thermodynamics. But it too fails to vindicate biological essentialism. What more, then, is required?

The key idea, I think, is that the membership condition must be *explanatory.* The essentialist hypothesizes that there exists some characteristic unique to and shared by all members of *Homo sapiens* which explains why they are the way they are. A species essence will be a causal mechanism that acts on each member of the species, making it the kind of thing that it is.

The characterization of essentialism just presented is fairly vague. For one thing, a great deal will depend on how one understands the crucial idea of *explanation.* But since explanation is clearly to be a scientific notion, I hope that, on my sketch, essentialism has the appearance of a scientific thesis, although perhaps one that is not terribly precise. Although historically prey to obscurantism, essentialism has nothing essentially to do with mystery mongering, or with the irrational injunction that one should ignore empirical data. It is a perfectly respectable claim about the existence of hidden structures which unite diverse individuals into natural kinds.

Besides its stress on the importance of explanations, there is another feature of our characterization of essentialism which will be important in what follows. The essentialist requires that a *species* be defined in terms of the characteristics of the *organisms* which belong to it. We might call this kind of definition a *constituent definition;* wholes are to be defined in terms of their parts, sets are to be defined in terms of their members, and so on. Pre-Darwinian critics of the species concept, like Buffon and Bonnet, argued that species are *un*real, because no such characteristics of organisms can be singled out (see Lovejoy 1936), and pre-Darwinian defenders of the species concept likewise agreed that the concept is legitimate only if constituent definitions could be provided. Constituent definitions are *reductionistic,* in that concepts at higher levels of organization (e.g., species) are legitimate only if they are definable in terms of concepts applying at lower levels of organization (e.g., organisms). It is quite clear that if there are finitely many levels of organization, one cannot demand constituent definitions for concepts at *every* level of organization (Kripke 1978). As we will see in what follows, evolutionary theory emancipated the species concept from the requirement that it be provided with a constituent definition. The scientific coherence of discourse at the population level was to be assured in another way, one to which the label "population thinking" is especially appropriate.

Chemistry is *prima facie* a discipline in which essentialist thinking has been vindicated. The periodic table of elements is a taxonomy of chemical kinds. The essence of each kind is its atomic number. Not only is it the case that all actual samples of nitrogen happen to have atomic number 14; it is necessarily the case that a thing is made of nitrogen if and only if it is made of stuff having atomic number 14. Moreover, this characteristic atomic number plays a central role in explaining other chemical properties of nitrogen. Although things made of this substance differ from each other in numerous respects, underlying this diversity there is a common feature. It was hardly irrational for chemists to search for this feature, and the working assumption that such essences were out there to be found, far from stifling inquiry, was a principal contributor to that inquiry's bearing fruit.

Can an equally strong case be made for an essentialist view of biological species? One often hears it said that evolution undermined essentialism because the essentialist held that species are static, but from 1859 on we had conclusive evidence that species evolve. This comment makes a straw man of essentialism and is in any case historically untrue to the thinking of many essentialists. For one thing, notice that the discovery of the transmutation of elements has not in the slightest degree undermined the periodic table. The fact that nitrogen can be changed into oxygen does not in any way show that nitrogen and oxygen lack essences. To be nitrogen is to have one atomic number; to be oxygen is to have another. To change from nitrogen into oxygen, a thing must therefore shift from one atomic number to another. The mere fact of evolution does not show that species lack essences.

As an historical matter, some essentialists, like Agassiz (1859), did assert a connection between essentialism and stasis. But others considered the possibility that new species should have arisen on Earth since the beginning (if they thought that there was a beginning). Thus, Linnaeus originally hypothesized that all species were created once and for all at the beginning, but later in his career he changed his mind because he thought that he had discovered a species, *Peloria,* which arose through cross-species hybridization (Rabel 1939, Ramsbottom 1938). And in *Generation of Animals* (II 746a30), Aristotle himself speculates about the possibility of new species arising as fertile hybrids. Countenancing such species need have no effect on binomial nomenclature or on deciding which characteristics of organisms to view as diagnostic. The question of *when* various kinds emerged seems to be quite independent of *what* makes for differences between kinds.

Another, more plausible, suggestion concerning how evolution undermined essentialism, is this: The fact that species evolve *gradually* entails that the boundaries between species are vague. The essentialist holds that there are characteristics that all and only the members of a given species possess. But this is no longer a tenable view; it is just as implausible as demanding that there should be a precise number of dollars which marks the boundary between rich and poor. This is the Sorites problem. Since ancient Greece, we have known that being a heap of stones, being bald, and being rich are concepts beset by line-drawing problems. But, the suggestion goes, it was only since 1859 that we have come to see that *Homo sapiens* is in the same boat. For example, Hull (1965) has argued that essentialism was refuted because of its Aristotelian theory of *definition;* the requirement that species have nontrivial necessary and sufficient conditions runs afoul of the kind of continuity found in nature.

Unfortunately, this limpid solution to our problem becomes clouded a bit when we consider the historical fact that many essentialists conceded the existence of line-drawing problems. Thus, Aristotle in his *History of Animals* (5888b4 ff.) remarks:

> . . . nature proceeds little by little from inanimate things to living creatures, in such a way that we are unable, in the continuous sequence to determine the boundary line between them or to say on which side an intermediate kind falls. Next, after inanimate things come the plants: and among the plants there are differences between one kind and another in the extent to which they seem to share in life, and the whole genus of plants appears to be alive when compared with other objects, but seems lifeless when compared with animals. The transition from them to the animals is a continuous one, as remarked before. For with some kinds of things found in the sea one would be at a loss to tell whether they are animals or plants.

It is unclear exactly how one should interpret this remark. Does it indicate that there are in fact no boundaries in nature, or does it mean that boundaries are difficult to discern? From the time of Aristotle up to the time of Darwin, the principle of continuity seems to have coexisted peacefully with strong essentialist convictions in the minds of many thinkers (Lovejoy 1936). Bonnet, Akenside, and Robinet are eighteenth-century biologists who exemplify this curious combination of doctrines. Does this coexistence imply that the two doctrines are in fact compatible, or rather, does it show that their conceptual dissonance was a long time in being appreciated? To answer this question, let us return to our analogy with the transmutation of elements.

In what sense are the boundaries between chemical kinds any more definite than those which we encounter in biology? At first glance, there appears to be all the difference in the world: In the periodic table, we have discrete jumps – between atomic number 36 and atomic number 37 there are no intermediate atomic numbers to blur distinctions. But let us reflect for a moment on the mechanism of transmutation. Consider, as an example, the experiment which settled the question of how nitrogen can be transmuted into oxygen (Ihde 1964, p. 509):

$$^{4}_{2}\mathrm{He} + ^{14}_{7}\mathrm{N} \rightarrow ^{17}_{8}\mathrm{O} + ^{1}_{1}\mathrm{H}.$$

In this reaction, the α-particle is absorbed and a proton is expelled. Let us ask of this process a typical Sorites question: At what point does the bombarded nucleus cease to be a nitrogen nucleus and when does it start being a nucleus of oxygen?

There *may* be a precise and principled answer to this question which is given by the relevant physical theory. But then again there may not.[2] I would suggest that which of these outcomes prevails really does not matter to the question of whether essentialism is a correct doctrine concerning the chemical kinds. It well may be that having a particular atomic number is a vague concept. But this is quite consistent with that (vague) property's being the essence of a chemical kind. This really does not matter, as long as the vagueness of "nitrogen" and that of "atomic number 14" coincide. Essentialism is in principle consistent with *vague essences.*[3] In spite of this, one wonders what the history of chemistry, and its attendant metaphysics, would have looked like, if the transmutation of elements had been a familiar phenomenon during the second half of the nineteenth century. Just as the fact of evolution at times tempted Darwin to adopt a nominalist attitude toward species,[4] so in chemistry the impressive taxonomy which we now have in the form of the periodic table might never have been developed, line-drawing problems having convinced chemists that chemical kinds are unreal.

As a historical matter, Hull (1965) was right in arguing that essentialism was standardly associated with a theory of definition in which vagueness is proscribed. Given this association, nonsaltative evolution was a profound embarrassment to the essentialist. But, if I am right, this theory of definition is inessential to essentialism. The argument that the gradualness of evolution is not the decisive issue in undermining essentialism is further supported, I think, by the fact that contemporary evolutionary theory contains proposals in which evolutionary gradualism is rejected. Eldredge and Gould (1972) have argued that the standard view of speciation (as given, for example, in Ayala 1978 and Mayr 1963) is one in which phylogeny is to be seen as a series of "punctuated equilibria." Discontinuities in the fossil record are not to be chalked up to incompleteness, but rather to the fact that, in geological time, jumps are the norm. I would suggest that this theory of discontinuous speciation is cold comfort to the essentialist. Whether lines are easy or hard to draw is not the main issue, or so I shall argue.[5]

Another local anti-essentialist argument has been developed by Ghiselin (1966, 1969, 1974) and Hull (1976, 1978). They have argued that evolutionary theory makes it more plausible to view species as spatio-temporally extended individuals than as natural kinds. A genuine natural kind like gold may "go extinct" and then reappear; it is quite possible for there to be gold things at one time, for there to be no gold at some later time, and then, finally, for gold to exist at some still later time. But the conception of species given by evolutionary theory does not allow this sort of flip-flopping in and out of existence: Once a biological taxon goes extinct, it must remain so. Hull (1978) argues that the difference between chemical natural kinds and biological species is that the latter, but not the former, are historical entities. Like organisms, biological species are individuated in part by historical criteria of spatio-temporal continuity. I am inclined to agree with this interpretation; its impact on pre-Darwinian conceptions of species could hardly be more profound. But what of its impact on essentialism? If essentialism is simply the view that species have essential properties (where a property need not be purely qualitative), then the doctrine remains untouched (as Hull himself grants). Kripke (1972) has suggested that each individual human being has the essential property of being born of precisely the sperm and the egg of which he or she was born. If such individuals as organisms have essential properties, then it will presumably also be possible for individuals like *Drosophila melanogaster* to have essential properties as well. Of course, these essences will be a far cry from the "purely qualitative" characteristics which traditional essentialism thought it was in the business of discovering.

My analysis of the impact of evolutionary theory on essentialism is parallel, though additional. Whether species are natural kinds or spatio-temporally extended individuals, essentialist theories about them are untenable. Two lines of argument will be developed for this conclusion. First, I will describe the way in which essentialism seeks to explain the existence of variability, and will argue that this conception is rendered implausible by evolutionary theory. Second, I will show how evolutionary theory has removed *the need* for providing species with constituent definitions; population thinking provides another way to make species scientifically intelligible. This consideration, coupled with the principle of parsimony, provides an additional reason for thinking that species do not have essences.

2. ARISTOTLE'S NATURAL STATE MODEL

One of the fundamental ideas in Aristotle's scientific thinking is what I will call his "Natural State Model." This model provides a technique for explaining the great diversity found in natural objects. Within the domain of physics, there are heavy and light objects, ones that move violently and ones that do not move at all. How is one to find some order that unites and underlies all this variety? Aristotle's hypothesis was that there is a distinction between the *natural state* of a kind of object and those states which are not natural. These latter are produced by subjecting the object to an *interfering force.* In the sublunar sphere, for a heavy object to be in its natural state is for it to be located where the center of the Earth is now (*On the Heavens,* ii, clr, 296b and 310b, 2–5). But, of course, many heavy objects fail to be there. The cause for this divergence from what is natural is that these objects are acted on by interfering forces that prevent them from achieving their natural state by frustrating their natural tendency. Variability within nature is thus to be explained as a deviation from what is natural; were there no interfering forces, all heavy objects would be located in the same place (Lloyd 1968).

Newton made use of Aristotle's distinction, but disagreed with him about what the natural state of physical objects is. The first law of motion says that if a body is not acted upon by a force, then it will remain at rest or in uniform motion. And even in general relativity, the geometry of space-time specifies a set of geodesics along which an object will move as long as it is not subjected to a force. Although the terms "natural" and "unnatural" no longer survive in Newtonian and post-Newtonian physics, Aristotle's distinction can be drawn within those theories. If there are no forces at all acting on an object, then, *a fortiori,* there are no interfering forces acting on it either. A natural state, within these theories, is a zero-force state.

The explanatory value of Aristotle's distinction is fairly familiar. If an object is not in its natural state, we know that the object must have been acted on by a force, and we set about finding it. We do this by consulting our catalog of known forces. If none of these is present, we might augment our catalog, or perhaps revise our conception of what the natural state of the system is. This pattern of analysis is used in population genetics under the rubric of the Hardy–Weinberg law. This law specifies an equilibrium state for the frequencies of genotypes in a panmictic population; this natural state is achieved when the evolutionary forces of mutation, migration, selection, and drift are not at work.

In the biological world, Aristotle sets forth the same sort of explanatory model. Diversity was to be accounted for as the joint product of natural tendencies and interfering forces. Aristotle invokes this model when he specifies the regularities governing how organisms reproduce themselves:

> . . . [for] any living thing that has reached its normal development and which is unmutilated, and whose mode of generation is not spontaneous, the most natural act is the production of another like itself, an animal producing an animal, a plant a plant . . . (*De Anima,* 415a26).

Like producing like, excepting the case of spontaneous generation, is the natural state, subject to a multitude of interferences, as we shall see.

In the case of spontaneous generation, the natural state of an organism is different. Although in the *Metaphysics* and the *Physics* "spontaneous" is used to mean unusual or random, in the later biological writings, *History of Animals* and *Generation of Animals,* Aristotle uses the term in a different way (Balme 1962, Hull 1967). Spontaneous generation obeys its own laws. For a whole range of organisms classified between the intermediate animals and the plants, like *never* naturally produces like. Rather, a bit of earth will spontaneously generate an earthworm, and the earthworm will then produce an eel. Similarly, the progression from slime to ascarid to gnat and that from cabbage leaf to grub to caterpillar to chrysallis to butterfly likewise counts as the natural reproductive pattern for this part of the living world (*History of Animals,* 570a5, 551b26, 551a13).

So much for the natural states. What counts as an interference for Aristotle? According to Aristotle's theory of sexual reproduction, the male semen provides a set of instructions which dictates how the female matter is to be shaped into an organism.[6] Interference may arise when the form fails to completely master the matter. This may happen, for example, when one or both parents are abnormal, or when the parents are from different species, or when there is trauma during foetal development. Such interferences are anything but rare, according to Aristotle. Mules – sterile hybrids – count as deviations from the natural state (*Generation of Animals,* ii, 8). In fact, the females of a species do too, even though they are necessary for the species to reproduce itself

(*Generation of Animals,* ii, 732a; ii, 3, 737a27; iv, 3, 767b8; iv, 6, 775a15). In fact, reproduction that is completely free of interference would result in an offspring who exactly resembles the father.[7] So failure to exactly resemble the male parent counts as a departure from the natural state. Deviations from type, whether mild or extreme, Aristotle labels "*terata*" – monsters. They are the result of interfering forces (*biaion*) deflecting reproduction from its natural pattern.

Besides trying to account for variation within species by using the Natural State Model, Aristotle at times seems to suggest that there are entire species that count as monsters (Preuss 1975, pp. 215–16; Hull 1968). Seals are deformed as a group because they resemble lower classes of animals, owing to their lack of ears. Snails, since they move like animals with their feet cut off, and lobsters, because they use their claws for locomotion, are likewise to be counted as monsters (*Generation of Animals,* 19, 714b, 18–19; *Parts of Animals,* iv, 8, 684a35). These so-called "dualizing species" arise because they are the best possible organisms that can result from the matter out of which they are made. The scale of nature, it is suggested, arises in all its graduated diversity because the quality of the matter from which organisms are made also varies – and nature persists in doing the best possible, given the ingredients at hand.

One cannot fault Aristotle for viewing so much of the biological domain as monstrous. Natural state models habitually have this characteristic; Newton's first law of motion is not impugned by the fact that no physical object is wholly unaffected by a force. Even so, Aristotle's partition of natural state and non-natural state in biology sounds to the modern ear like a reasonable distinction run wild. "Real terata are one thing," one might say, "but to call entire species, and all females, and all males who don't exactly resemble their fathers monsters, seems absurd." Notice that our "modern" conceptions of health and disease and our notion of normality as something other than a statistical average enshrine Aristotle's model. We therefore are tempted to make only a conservative criticism of Aristotle's biology: We preserve the form of model he propounded, but criticize the applications he made of it. Whether this minimal critique of Aristotle is possible in the light of evolutionary theory remains to be seen.

The Natural State Model constitutes a powerful tool for accounting for variation. Even when two species seem to blend into each other continuously, it may still be the case that all the members of one species have one natural tendency while the members of the other species have a quite different natural tendency. Interfering forces may, in varying degrees, deflect the individuals in both species from their natural states, thus yielding the surface impression that there are no boundaries between the species. This essentialist response to the fact of diversity has the virtue that it avoids the *ad hoc* maneuver of contracting the boundaries of species so as to preserve their internal homogeneity.[8] This latter strategy was not unknown to the essentialist, but its methodological defects are too well known to be worth recounting here. Instead of insisting that species be defined in terms of some surface morphological feature, and thereby having each species shrink to a point, the essentialist can countenance unlimited variety in, and continuity between, species, as long as underlying this plenum one can expect to find discrete natural tendencies. The failure to discover such underlying mechanisms is not a conclusive reason to think that none exist; but the development of a theory that implies that natural tendencies are not part of the natural order is another matter entirely.

Aristotle's model was a fixed point in the diverse conjectures to be found in pre-Darwinian biology. Preformationists and epigeneticists, advocates of evolution and proponents of stasis, all assumed that there is a real difference between natural states and states caused by interfering forces. The study of monstrosity – teratology – which in this period made the transition from unbridled speculation to encyclopedic catalogues of experimental oddities (Meyer 1939), is an especially revealing example of the power exerted by the Natural State Model. Consider, for example, the eighteenth-century disagreement between Maupertuis and Bonnet over the proper explanation of polydactyly. Both had at their fingertips a genealogy; it was clear to both that somehow or other the trait regularly reappeared through the generations. Maupertuis conjectured that defective hereditary material was passed along, having originally made its appearance in the family because of *an error in nature* (Glass 1959b, pp. 62–7). Maupertuis, a convinced Newtonian, thought that traits, both normal and anomalous, resulted from the lawful combination of hereditary particles (Roger 1963). When such particles have normal quantities of attraction for each other, normal characteristics result. However, when particles depart from this natural state, either too many or too few of them combine, thus resulting in *monstres par exces* or *monstres par defaut.* Bonnet, a convinced ovist, offered a different hypothesis. For him, polydactyly is never encoded in the germ, but rather results from abnormal interuterine conditions or from male sperm interfering with normal development (Glass 1959a, p. 169). Thus whether polydactyly is "naturalized" by Maupertuis' appeal to heredity or by Bonnet's appeal to environment, the trait is never regarded as being completely natural. Variability in nature – in this case variability as to the number of digits – is a deviation from type.

In pre-Darwinian disputes over evolution, natural states loom equally large. Evolutionary claims during this period mainly assumed that living things were programmed to develop in a certain sequence, and that the emergence of biological novelty was therefore in conformity with some natural plan. Lovejoy (1936) discusses how the Great Chain of Being was "temporalized" during the eighteenth century; by this, he has in mind the tendency to think that the natural ordering of living things from those of higher type down to those of lower type also represented an historical progression. Such programmed, directed evolution – in which some types naturally give rise to others – is very much in the spirit of the Natural State Model. Whether species are subject to historical unfolding, or rather exist unchanged for all time, the concept of species was inevitably associated with that of type; on either view, variation is deviation caused by interfering forces.

It was generally presupposed that somewhere within the possible variations that a species is capable of, there is a privileged state – a state which has a special causal and explanatory role. The laws governing a species will specify this state, just as the laws that make sense of the diversity of kinematic states found in physics tell us which is the natural state of a physical object. The diversity of individual organisms is a veil which must be penetrated in the search for invariance. The transformation in thinking which we will trace in the next two sections consisted in the realization that this diversity itself constituted an invariance, obeying its own laws.

3. THE LAW OF ERRORS AND THE EMERGENCE OF POPULATION THINKING

So far, I have sketched several of the applications that have been made of Aristotle's model within biology. This strategy for explaining variation, I will argue in the next section, has been discredited by modern evolutionary theory. Our current theories of biological variation provide no more role for the idea of natural state than our current physical theories do for the notion of absolute simultaneity. Theories in population genetics enshrine a different model of variation, one which emerged during the second half of the nineteenth century. Some brief account of the evolution within the field of statistics of our understanding of the law of errors will lay the groundwork for discussing the modern understanding of biological variation.

From its theoretical formulation and articulation in the eighteenth century, up until the middle of the nineteenth century, the law of errors was understood as a law about *errors.* Daniel Bernouilli, Lagrange, and Laplace each tried to develop mathematical techniques for determining how a set of discordant observations was to be interpreted (Todhunter 1865). The model for this problem was, of course, that there is a single true value for some observational variable, and a multiplicity of inconsistent readings that have been obtained. Here we have a straightforward instance of Aristotle's model: Interfering forces cause variation in opinion; in nature there is but one true value. The problem for the theory of errors was to penetrate the veil of variability and to discover behind it the single value which was the constant cause of the multiplicity of different readings. Each observation was thus viewed as the causal upshot of two kinds of factors: Part of what determines an observational outcome is the real value of the variable, but interfering forces which distort the communication of this information from nature to mind also play a role. If these interfering forces are random – if they are as likely to take one value as any other – then the mean value of the readings is likely to represent the truth, when the number of observations is large. In this case, one reaches the truth by ascending to the summit of the bell curve. It is important to notice that this application of the Natural State Model is epistemological, not ontological. One seeks to account for variation in our observations of nature, not variation in nature itself. The decisive transition, from this epistemological to an ontological application, was made in the 1830s by the influential Belgian statistician Adolphe Quetelet.

Quetelet's insight was that the law of errors could be given an ontological interpretation by invoking a distinction which Laplace had earlier exploited in his work in Newtonian mechanics.[9] Laplace decomposed the forces at work in the solar system into two kinds. First, there are the *constant causes* by which the planets are affected by the sun's gravitation; second, there are the particular *disturbing causes* which arise from the mutual influences of the planets, their satellites, and the comets. Laplace's strategy was a familiar analytic one. He tried to decompose the factors at work in a phenomenon into components, and to analyze their separate contributions to the outcome. The character of this decomposition, however, is of special interest: One central, causal agent is at work on the components of a system, but the effects of this force are complicated by the presence of numerous interferences which act in different directions.

In his book of 1835, *Sur l'homme et le développement de ses facultés, ou essai de physique social,* Quetelet put forward his conception of the *average man,* which for him constituted the true subject of the discipline of social physics. By studying the average man, Quetelet hoped to filter out the multifarious and idiosyncratic characteristics which make for diversity in a population, and to focus on the central facts which constitute the social body itself. Like Weber's later idea of an ideal type, Quetelet's conception of the average man was introduced as a "fiction" whose utility was to facilitate a clear view of social facts by allowing one to abstract from the vagaries of individual differences. But unlike Weber, Quetelet quickly came to view his construct as real – a subject matter in its own right. Quetelet was struck by the analogy between a society's average man and a physical system's center of gravity. Since the latter could play a causal role, so too could the former; neither was a mere abstraction. For Quetelet, variability within a population *is caused by* deviation from type. When the astronomer John Herschel reviewed Quetelet's *Lettres sur les probabilités* in 1850, he nicely captured Quetelet's idea that the average man is no mere artefact of reflection:

> An average may exist of the most different objects, as the heights of houses in a town, or the sizes of books in a library. It may be convenient to convey a general notion of the things averaged; but it involves no conception of a natural and recognizable central magnitude, all differences from which ought to be regarded as deviations from a standard. The notion of a mean, on the other hand, does imply such a conception, standing distinguished from an average by this very feature, *viz.* the regular marching of the groups, increasing to a maximum and thence again diminishing. An average gives us no assurance that the future will be like the past. A mean may be reckoned on with the most implicit confidence. (Hilts 1973, p. 217)

Quetelet found little theoretical significance in the fact of individual differences. Concepts of correlation and amount of variation were unknown to him. For Quetelet, the law of errors is still a law about errors, only for him the mistakes are made by nature, not by observers. Our belief that there is variation in a population is no mistake on our part. Rather, it is the result of interferences confounding the expression of a prototype. Were interfering forces not to occur, there would be no variation.

It may strike the modern reader as incredible that anyone could view a trait like girth on this mode. However, Quetelet, who was perhaps the most influential statistician of his time, did understand biological difference in this way. He was impressed, not to say awestruck, by the fact that the results of accurately measuring the waists of a thousand Scottish soldiers would assume the same bell-shaped distribution as the results of inaccurately measuring the girth of a single, average, soldier a thousand times. For Quetelet, the point of attending to variation was to *see through it* – to render it transparent. Averages were the very antitheses of artefacts; they alone were the true objects of inquiry.[10]

Francis Galton, who was Darwin's cousin,[11] was responsible for fundamental innovations in the analysis of individual differences. He discovered the standard deviation and the correlation coefficient. His work on heredity was later claimed by both Mendelians and biometricians as seminal, and thus can be viewed as a crucial step toward the synthetic theory of evolution (Provine 1971). But his interest to our story is more restricted. Galton, despite his frequently sympathetic comments about the concept of type,[12] helped to displace the average man and the idea of deviation from type. He did this, not by attacking these typological constructs directly, but by developing an alternative model for accounting for variability. This model is a nascent form of the kind of population thinking in which evolutionary biologists today engage.

One of Galton's main intellectual goals was to show that heredity is a central cause of individual differences. Although the arguments which Galton put forward for his hereditarian thesis were weak, the conception of variability he exploited in his book *Hereditary Genius* (1869) is of great significance. For Galton, variability is *not* to be explained away as the result of interference with a single prototype. Rather, variability within one generation is explained by appeal to variability in the previous generation and to facts about the transmission of variability. Galton used the law of errors, but no longer viewed it as a law *about* errors. As Hilts (1973, pp. 223–4) remarks: "Because Galton was able to associate the error distribution with individual differences caused by heredity, the distinction between constant and accidental causes lost much of its meaning." At the end of his life, Galton judged that one of his most important ideas was that the science of heredity should be concerned with deviations measured in statistical units. Quetelet had earlier denied that such units exist. Galton's discovery of the standard deviation gave him the mathematical machinery to begin treating variability as obeying its own laws, as something other than an idiosyncratic artefact.

Eight years after the publication of *Hereditary Genius,* Galton was able to sketch a solution for the problem he had noted in that work: What fraction of the parental deviations from the norm are passed on to offspring? Galton described a model in which hereditary causes and non-hereditary causes are partitioned. Were only the former of these at work, he conjectured, each child would have traits that are intermediate between those of its parents. In this case, the amount of variation would decrease in each generation. But Galton suspected that the amount of variation is constant across generations. To account for this, he posited a second, counteracting force which causes variability within each family. Were this second force the only one at work, the amount of variation would increase. But in reality, the centrifugal and centripetal forces combine to yield a constant quantity of variability across the generations. An error distribution is thus accounted for by way of a hypothesis which characterizes it as the sum of two other error distributions.

In his *Natural Inheritance* of 1889, Galton went on to complete his investigations of the correlation coefficient, and introduced the name "normal law" as a more appropriate label for what had traditionally been called the law of errors.[13] Bell curves are normal; they are found everywhere, Galton thought. This change in nomenclature crystallized a significant transformation in thinking. Bell curves need not represent mistakes made by fallible observers or by sportive nature. Regardless of the underlying etiology, *they are real;* they enter into explanations because the variability they represent is lawful and causally efficacious.

The transition made possible by statistical thinking from typological to population thinking was not completed by Galton.[14] Although his innovations loosened the grip of essentialism, he himself was deeply committed to the idea of racial types and believed that evolutionary theory presupposes the reality of types. Both Galton and Darwin (1859, ch. 5; 1868, ch. 13) spoke sympathetically about the ideas of unity of type and of reversion to type, and sought to provide historical justifications of these ideas in terms of common descent. Unity of type was just similarity owing to common ancestry; reversion to type was the reappearance of latent ancestral traits. But the presence of these ideas in their writings should not obscure the way in which their theorizing began to undermine typological thinking.

Darwin and Galton focused on the population as a unit of organization. The population is an entity, subject to its own forces, and obeying its own laws. The details concerning the individuals who are parts of this whole are pretty much irrelevant. Describing a single individual is as theoretically peripheral to a populationist as describing the motion of a single molecule is to the kinetic theory of gases. In this important sense, population thinking involves *ignoring individuals:* It is holistic, not atomistic. This conclusion contradicts Mayr's (1959, p. 28) assertion that for the populationist, "the individual alone is real."

Typologists and populationists agree that averages exist, and both grant the existence of variation. They disagree about the explanatory character of these. For Quetelet, and for typologists generally, variability does not explain anything. Rather it is something to be explained or explained away. Quetelet posited a process in which uniformity gives rise to diversity; a single prototype – the average man – is mapped onto a variable resulting population. Galton, on the other hand, explained diversity in terms of an earlier diversity and constructed the mathematical tools to make this kind of analysis possible.

Both typologists and populationists seek to transcend the blooming, buzzing confusion of individual variation. Like all scientists, they do this by trying to identify properties of systems that remain constant in spite of the system's changes. For the typologist, the search for invariances takes the form of a search for natural tendencies. The typologist formulates a causal hypothesis about the forces at work on each individual within a population. The invariance underlying this diversity is the possession of a particular natural tendency *by each individual organism.* The populationist, on the other hand, tries to identify invariances by ascending to a different level of organization. For Galton, the invariant property across generations within a lineage is the amount of variability, and this is a property *of populations.* Again we see a way in which the essentialist is more concerned with individual organisms than the populationist is. Far from ignoring individuals, the typologist, *via* his use of the Natural State Model, resolutely focuses on individual organisms as the entities that possess invariant properties. The populationist, on the other hand, sees that it is not just individual organisms that can be the bearers of unchanging characteristics. Rather than looking for a reality that *underlies* diversity, the populationist can postulate a reality *sustained* by diversity.

I have just argued that there is an important sense in which typologists are more concerned with individual organisms than populationists are. However, looked at in another way, Mayr's point that populationists assign a more central role to organisms than typologists do can be established. In models of natural selection in which organisms enjoy different rates of reproductive success because of differences in fitness, natural selection is a force that acts on individual (organismic) differences. This Darwinian view assigns a causal role to individual idiosyncrasies. Individual differences are not *the effects* of interfering forces confounding the expression of a prototype; rather they are *the causes* of events that are absolutely central to the history of evolution. It is in this sense that Mayr is right in saying that evolutionary theory treats individuals as real in a way that typological thought does not (see also Lewontin 1974, pp. 5–6). Putting my point and Mayr's point, thus interpreted, together, we might say that population thinking endows individual organisms with more reality *and* with less reality than typological thinking attributes to them.

To be real is to have causal efficacy; to be unreal is to be a mere artefact of some causal process. This characterization of what it is to be real, also used by Hacking (1975), is markedly different from the one used in traditional metaphysical disputes concerning realism, verificationism, and idealism (Sober 1982). There, the problem is not how things are causally related, but rather it concerns what in fact *exists,* and whether what exists exists "independently" of us. The causal view of what it is to be real offers an explanation of a peculiar fact that is part of the more traditional metaphysical problem. Although two predicates may name real physical properties, natural kinds, theoretical magnitudes, or physical objects, simple operations on that pair of predicates may yield predicates which fail to name anything real. Thus, for example, "mass" and "charge" may name real physical magnitudes, even though "$mass^2/charge^3$" fails to name anything real. This is hard to explain, if reality is simply equated with existence (or with existence-that-is-independent-of-us). After all, if an object has a mass and if it has a charge, then there must be such a thing as what the square of its mass over the cube of its charge is. While this is quite true, it is *not* similarly correct to infer that because an object's mass causes some things and its charge causes other things, then there must be something which is caused by the square of its mass divided by the cube of its charge. Realism, in this case at least, is a thesis about what is cause and what is effect.

If we look forward in time, from the time of Galton and Darwin to the Modern Synthesis and beyond, we can see how population models have come to play a profoundly important role in evolutionary theorizing. In such models, properties of populations are identified and laws are formulated about their interrelations. Hypotheses in theoretical ecology and in island biogeography, for example, *generalize over populations* (see, for example, Wilson and Bossert 1971, chs. 3 and 4). The use of population concepts is not legitimized in those disciplines by defining them in terms of concepts applying at some lower level of organization. Rather, the use of one population concept is vindicated by showing how it stands in law-like relations with other concepts *at the same level of organization.* It is in this way that we can see that there is an alternative to constituent definition. Here, then, is one way in which evolutionary theorizing undermined essentialism: Essentialism requires that species concepts be legitimized by constituent definition, but evolutionary theory, in its articulation of population models, makes such demands unnecessary. Explanations can proceed without this reductionistic requirement being met.

If this argument is correct, there is a standard assumption made in traditional metaphysical problems having to do with identity which needs to be reevaluated. There could hardly be a more central category in our metaphysics, both scientific and everyday, than that of an enduring physical object. The way philosophers have tried to understand this category is as follows: Imagine a collection of instantaneous objects – i.e., objects at a moment in time. How are these various instantaneous objects united into the temporally enduring objects of our ontology? What criteria do we use when we lump together some time slices, but not others? This approach to the problem is basically that of looking for a constituent definition: Enduring objects are to be defined out of their constituent time-slices. But if populations can be scientifically legitimized in ways other than by using constituent definitions, perhaps the same thing is true of the category of physical object itself. I take it that Quine's (1953a) slogan "no entity without identity" is basically a demand for constituent definitions; this demand, which may be fruitful in some contexts, should not be generalized into a universal maxim (nor can it be, if there are finitely many levels of organization. (See Kripke 1978).

4. THE DISAPPEARANCE OF A DISTINCTION

The fate of Aristotle's model at the hands of population biology bears a striking resemblance to what happened to the notion of absolute simultaneity with the advent of relativity theory. Within classical physics, there was a single, well-defined answer to the question "What is the temporal separation of two events *x* and *y*?" However, relativity theory revealed that the answer to this question depends on one's choice of a rest frame; given different rest frames, one gets different answers. As is well known, the classical notions of temporal separation and spatial separation gave way in relativity theory to a magnitude that is not relative at all: This is the spatio-temporal separation of the two events. How large this quantity is does not depend on any choice of rest frame; it is frame invariant. Minkowski (1908) took this fact about relativity theory to indicate that space and time are not real physical properties at all, since they depend for their values on choices that are wholly arbitrary. For Minkowski, to be real is to be invariant, and space and time become mere shadows.

Special relativity fails to discriminate between the various values that may be assigned to the temporal interval separating a pair of events; they are all on a par. No one specification of the temporal separation is any more correct than any other. It would be utterly implausible to interpret this fact as indicating that there is a physically real distinction which special relativity fails to make. The fact that our best theory fails to draw this distinction gives us a very good reason for suspecting that the distinction is unreal, and this is the standard view of the matter which was crystallized in the work of Minkowski.

According to the Natural State Model, there is one path of foetal development which counts as the realization of the organism's natural state, while other developmental results are consequences of unnatural interferences. Put slightly differently, for a given genotype, there is a single phenotype which it can have that is the natural one. Or, more modestly, the requirement might be that there is some restricted range of phenotypes which count as natural. However, when one looks to genetic theory for a conception of the relation between genotype and phenotype, one finds no such distinction between natural state and states which are the results of interference. One finds, instead, the *norm of reaction,* which graphs the different phenotypic results that a genotype can have in different environments.[15] For example, the height of a single corn plant genotype might vary according to the ambient temperature. How would one answer the question: "Which of the phenotypes is the natural one for the corn plant to have?" One way to take this obscure question is indicated by the following answer: Each of the heights indicated in the norm of reaction is as "natural" as any other, since each happens in nature. Choose an environment, and relative to that choice we know what the phenotypic upshot in that environment is. But, of course, if the question we are considering is understood in terms of the Natural State Model, this sort of answer will not do. The Natural State Model presupposes that there is some phenotype which is the natural one *which is independent of a choice of environment.* The Natural State Model presupposes that there is some environment which is the natural environment for the genotype to be in, which determines, in conjunction with the norm of reaction, what the natural phenotype for the genotype is. But these presuppositions find no expression in the norm of reaction: All environments are on a par, and all phenotypes are on a par. The required distinctions simply are not made.

When one turns from the various phenotypes that a single genotype might produce, to the various genotypes that a population might contain, the same result obtains. Again, according to the Natural State Model, there is a single genotype or restricted class of genotypes, which count as the natural states of the population or species, all other genotypes being the result of interfering forces. But again, statistical profiles of genotypic variation within a population enshrine no such difference. Genotypes differ from each other in frequency; but unusual genotypes are not in any literal sense to be understood as deviations from type.

When a corn plant of a particular genotype withers and dies, owing to the absence of trace elements in the soil, the Natural State Model will view this as an outcome that is not natural. When it thrives and is reproductively successful, one wants to say that *this* environment might be the natural one. Given these ideas, one might try to vindicate the Natural State Model from a selectionist point of view by identifying the natural environment of a genotype with the environment in which it is fittest.[16]

This suggestion fails to coincide with important intuitions expressed in the Natural State Model. First of all, let us ask the question: What is the range of environments relative to which the fittest environment is to be understood? Shall we think of the natural state as that which obtains when the environment is the fittest *of all possible environments*? If so, the stud bull, injected with medications, its reproductive capacities boosted to phenomenal rates by an efficient artificial insemination program, has achieved its natural state. And in similar fashion, the kind of environment that biologists use to characterize the intrinsic rate of increase (r) of a population – one in which there is no disease, no predation, no limitations of space or food supplies – will likewise count as the natural environment. But these optimal environments are *not natural,* the Natural State Model tells us. They involve "artificially boosting" the fitness of resulting phenotypes by placing the genotypes in environments that are more advantageous than the natural environment.

Let us consider another, perhaps more plausible, way to understand the range of environments with respect to which the fittest environment is to be calculated. Instead of taking the best of all possible environments, why not, more modestly, consider the best of all environments that have been historically represented? This suggestion evades the second, but not the first, counterexample mentioned above. However, other problems present themselves. The natural state of a genotype is often understood to be one which has yet to occur. Perhaps every environment that a species has historically experienced is such that a given genotype in that environment results in a *diseased* phenotype, or one which is developmentally impaired in some way. The natural state of a genotype is often taken to be some sort of ideal state which may or may not be closely approximated in the history of the species.

I have just argued that the idea of a fittest environment does not allow one to impose on the norm of reaction the kind of distinction that the Natural State Model requires. Precisely the same reasons count against construing the idea of a genotype's being the natural state of a species in terms of maximal fitness. It is part of the Natural State Model that the natural genotype for a species can be less fit (in some range of environments) than the best of all possible genotypes. And the natural genotype can likewise fail to be historically represented.

Aristotle is typical of exponents of the Natural State Model in holding that variation is introduced into a population by virtue of interferences with normal sexual reproduction. Current understanding of the mechanisms of reproduction shows that precisely the opposite is the case. Even if one dismisses mutations as "unnatural interferences," the fact of genetic recombination looms large. Generally, the number of total genotypes that a gene pool can produce by recombination is the product of the number of diploid genotypes that can be constructed at each locus. For species like *Homo sapiens* and *Drosophila melanogaster,* the number of loci has been estimated to be about 10,000 or more. What this means is that the number of genotypes that can be generated by recombination is greater than the number of atoms in the visible universe (Wilson and Bossert 1971, p. 39). For species with this number of loci, even a single male and a single female can themselves reproduce a significant fraction of the variation found in a population from which they are drawn. All sorts of deleterious phenotypes may emerge from the recombination process initiated by a founder population.

The Natural State Model is a *causal,* and thereby an *historical, hypothesis.* The essentialist attempts to understand variation within a species as arising through a process of deviation from type. By tracing back the origins of this variability we discover the natural state of a species. To do this is to uncover that natural tendency possessed by each member of the species. But the science which describes the laws governing the historical origins of variation within species makes no appeal to such "natural tendencies." Rather, this frame invariant "natural tendency" – this property that an organism is supposed to have regardless of what environment it might be in – has been replaced by a frame relative property – namely, the phenotype that a genotype will produce *in a given environment.* The historical concept of a natural state is discredited in much the same way that the physical concept of absolute simultaneity was.

5. CONCLUSION

Tenacious, if not pig-headed, adherence to a research program is necessary, if the conceptual possibilities in the program are to be explored adequately. It was hardly irrational for nineteenth-century research on the chemical elements to persist in its assumption that chemical kinds exist and have essential properties. The same holds true for those who hold that species are natural kinds and have essential properties; repeated failure to turn up the postulated items may be interpreted as simply showing that inquiry has not proceeded far enough. Matters change, however, when theoretical reasons start to emerge that cast doubt on the existence claim. For example, if the existence claim is shown to be theoretically superfluous, that counts as one reason for suspecting that no such thing exists. In another vein, if the causal mechanism associated with the postulated entity is cast in doubt, that too poses problems for the plausibility of the existence claim. Our discussion of how population thinking emancipated biology from the need for constituent definitions of species is an argument of the first kind. Our examination of the theory of variation presupposed by essentialism is an argument of the second kind.

No phenotypic characteristic can be postulated as a species essence; the norm of reaction for each genotype shows that it is arbitrary to single out as privileged one phenotype as opposed to any other. Similar considerations show that no genotypic characteristic can be postulated as a species essence; the genetic variability found in sexual populations is prodigious and, again, there is no biologically plausible way to single out some genetic characteristics as natural while viewing others as the upshot of interfering forces. Even if a species were found in which some characteristic is shared by all and only the organisms that are in the species, this could not be counted as a species essence. Imagine, for example, that some novel form of life is created in the laboratory and subjected to some extreme form of stabilizing selection. If the number of organisms is kept small, it may turn out that the internal homogeneity of the species, as well as its distinctness from all other species, has been assured. However, the explanation of this phenomenon would be given in terms of the selection pressures acting on the population. If the universal property were a species essence, however, explaining why it is universal would be like explaining why all acids are proton donors, or why all bachelors are unmarried, or why all nitrogen has atomic number 14. These latter necessary truths, if they are explainable at all, are not explained by saying that some contingent causal force acted on acids, bachelors, or samples of nitrogen, thereby endowing them with the property in question. Characteristics possessed by all and only the extant members of a species, if such were to exist, would not be species essences. It is for this reason that hypotheses that reject evolutionary gradualism in no way support the claims of essentialism.

The essentialist hoped to sweep aside the veil of variability found within species by discovering some natural tendency which each individual in the species possesses. This natural tendency was to be a dispositional property which would be manifest, were interfering forces not at work. Heterogeneity is thus the result of a departure from the natural state. But, with the development of evolutionary theory, it turned out that no such property was available to the essentialist, and in fact our current model of variability radically differs from the essentialist's causal hypothesis about the origins of variability.

At the same time that evolutionary theory undermined the essentialist's model of variability, it also removed the need for discovering species essences. Characteristics of populations do not have to be defined in terms of characteristics of organisms for population concepts to be coherent and fruitful. Population biology attempts to formulate generalizations about kinds of populations. In spite of the fact that species cannot be precisely individuated in terms of their constituent organisms, species undergo evolutionary processes, and the character of such processes is what population biology attempts to describe. Laws generalizing over population will, of course, include the standard *ceteris paribus* rider: they will describe how various properties and magnitudes are related, as long as no other forces affect the system. At least one such law describes what happens when *no* evolutionary force is at work in a panmictic Mendelian population. This is the Hardy–Weinberg equilibrium law. This law describes an essential property – a property which is necessary for a population to be Mendelian. But, of course, such laws do not pick out *species'* essences. Perhaps essentialism can reemerge as a thesis, not about species, but about *kinds* of species. The Natural State Model arguably finds an application at that level of organization in that the Hardy–Weinberg zero-force state is distinguished from other possible population configurations.

The transposition of Aristotle's distinction is significant. The essentialist searched for a property *of individual organisms* which is invariant across the organisms in a species. The Hardy–Weinberg law and other more interesting population laws, on the other hand, identify properties of *populations* which are invariant across all populations of a certain kind. In this sense, essentialism pursued an individualistic (organismic) methodology, which population thinking supplants by specifying laws governing objects at a higher level of organization. From the individualistic (organismic) perspective assumed by essentialism, species are real only if they can be delimited in terms of membership conditions applying to individual organisms. But the populationist point of view made possible by evolutionary theory made such reductionistic demands unnecessary. Since populations and their properties are subject to their own invariances and have their own causal efficacy, it is no more reasonable to demand a species definition in terms of the properties of constituent organisms than it is to require organismic biology to postpone its inquiries until a criterion for sameness of organism is formulated in terms of relations among constituent cells. Essentialism lost its grip when populations came to be thought of as real. And the mark of this latter transformation in thought was the transposition of the search for invariances to a higher level of organization.

NOTES

1. If species are *individuals* – spatio-temporally extended lineages – as Ghiselin (1966, 1969, 1974) and Hull (1976, 1978) have argued, then we have our assurance of finitude. If, on the other hand, species are kinds of things, which may in principle be found anywhere in the universe at any time, then a slightly different argument is needed for the claim that the same species is overwhelmingly unlikely to have evolved twice. Such an argument is provided by considering the way in which speciation depends on the coincidence of a huge number of initial conditions. See Ayala (1978) for a summary of the received view of this matter.
2. It is arguable that quantum mechanical considerations show that the concept of being a nucleus with a particular atomic number is a vague one. Presumably, a collection of protons constitutes a nucleus when the strong force that causes them to attract each other overcomes their mutual electromagnetic repulsion. Whether this happens or not is a function of the distances between the protons. But *this* concept – that of "the" distance

between particles – is indeterminate. Hence, the question of whether something is or is not a nucleus with a particular atomic number can only be answered probabilistically.

3. It is probably a mistake to talk about concepts being vague *simpliciter*. Rather, one should formulate matters in terms of concepts being vague relative to a particular application. The issue of whether a concept is vague seems to reduce to the issue of whether there are cases in which it is indeterminate whether the concept applies or not. I would guess that practically every concept applying to physical objects is vague in this sense. Thus, even such concepts as "being two in number" are such that circumstances can be described in which it is indeterminate whether or not they apply to the objects in question.
4. Thus Darwin (1859, p. 52) says: "From these remarks it will be seen that I look at the term species, as one arbitrarily given for the sake of convenience to a set of individuals closely resembling each other, and that it does not essentially differ from the term variety, which is given to less distinct and more fluctuating forms. The term variety, again, in comparison with mere individual differences, is also applied arbitrarily, and for mere convenience sake."
5. I am not suggesting that Hull (1965) and others have misidentified the essence of essentialism and that their criticisms thereby fail to get to the heart of the matter. Essentialism, like most isms which evolve historically, probably does not even have an essence. Rather, I am trying to construe essentialism as a fairly flexible doctrine which, in at least some circumstances, can be seen to be quite consistent with the existence of insoluble line-drawing problems.
6. This characterization of Aristotle's view in terms of some information-bearing entity is not completely anachronistic, as Delbrück (1971) points out when he (in jest) suggests that Aristotle should receive a Nobel Prize for having discovered DNA.
7. In this discussion of Aristotle's view of *terata,* I have been much helped by Furth's (1975, section 11).
8. If one views Aristotle as excluding monstrous forms from membership in any species category, then one will have an extreme instance of this *ad hoc* strategy; *no* organism will belong to any species. Hull (1973, pp. 39–40) sees Aristotle and scholastic science as hopelessly committed to this futile strategy. However, on the view I would attribute to Aristotle, most, if not all, monstrous forms are members of the species from which they arose. They, like Newtonian particles which fail to be at rest or in uniform motion, fail to achieve their natural states because of identifiable causal forces.
9. Hilts (1973, pp. 209–10). My discussion of Quetelet and Galton in what follows leans heavily on Hilts (1973). It has a number of points in common with Hacking (1975).

10. Boring (1929, p. 477) brings out the Aristotelian teleology contained in Quetelet's ideas quite well when he characterizes Quetelet as holding that "We might regard such human variation as if it occurred when nature aimed at an ideal and missed by varying amounts."
11. Although Galton found *The Origin of Species* an encouragement to pursue his own ideas, he indicates that his interest in variation and inheritance were of long standing. See Hilts (1973, p. 220).
12. In his *Hereditary Genius,* Galton compared the development of species with a many-faceted spheroid tumbling over from one facet or stable equilibrium to another. See Provine (1971, pp. 14–15). This saltative process ensured unity of type. In spite of Galton's adherence to the idea of discontinuous evolution and certain other essentialist ideas (Lewontin 1974, p. 4), his innovations in population thinking were anti-essentialist in their consequences, or so I will argue.
13. Hilts (1973, p. 228). Walker (1929, p. 185) claims that the origin of the name "normal curve" is obscure. It occurs in Lexis and, she says, "it is not improbable that the term goes back to Quetelet." As natural and inevitable as Quetelet found his interpretation of the bell curve in terms of the Natural State Model, by the time Galton's *Natural Inheritance* appeared in 1889, there was growing sentiment that this interpretation was acceptable, if at all, only as a special case. Thus we find Galton, in that work (p. 58), saying that "the term Probable Error is absurd when applied to the subjects now in hand, such as Stature, Eye-colour, Artistic Faculty, or Disease." A year earlier, Venn, in his *The Logic of Chance* (p. 42), made a similar comment: "When we perform an operation ourselves with a clear consciousness of what we are aiming at, we may quite correctly speak of every deviation from this as being an error; but when Nature presents us with a group of objects of every kind, it is using a rather bold metaphor to speak in this case also of a law of error, as if she had been aiming at something all the time, and had like the rest of us missed her mark more or less in every instance." Quotations are drawn from Walker (1929, p. 53).
14. It would be important to trace the development of statistical ideas from Galton through Pearson and his circle to R. A. Fisher, and to see whether Pearson's positivistic convictions had the effect of further proscribing the idea of types on the grounds that it is "unscientific." Cohen (1972) sees Galton as already adopting some positivistic attitudes in his idea that heredity was to be understood in terms of correlations, and not in terms of causal forces. Also, see Hacking (1975) for a bold attempt to link Galton's innovations to other developments in nineteenth-century thought. I should point out that a fuller treatment of the emergence of population thinking would have to ascribe a central role to Mendel. He, much more than Galton, provided the central elements of our present conception of the relation of heredity and variation. I have stressed Galton, however,

because of his interpretation of statistics and because of his view of the population as a unit of explanation.

15. The discussion of the norm of reaction in what follows depends heavily on some points made in Lewontin (1977).
16. This selectionist suggestion needs to be made more precise by specifying the notion of fitness used. I will not lay out these different conceptions here. Rather, I invite the reader to choose the one that he or she finds most plausible. The upshot of my argument does not seem to depend on which biologically plausible characterization is chosen.

REFERENCES

Agassiz, L. (1859). *Essay on Classification.* Cambridge, Mass.: Harvard University Press.

Ayala, F. (1978). "The Mechanisms of Evolution," *Scientific American 239, 3:* 56–69.

Balme, D. (1962). "Development of Biology in Aristotle and Theophrastus: Theory of Spontaneous Generation," *Phronesis, 2, 1:* 91–104.

Boring, E. (1929). *A History of Experimental Psychology.* New York: Appleton-Century-Crofts.

Cohen, R. (1972). "Francis Galton's Contribution to Genetics," *Journal of the History of Biology 5, 2:* 389–412.

Darwin, C. (1859). *On the Origin of Species.* Cambridge, Mass.: Harvard University Press.

Delbrück, M. (1971). "Aristotle-totle-totle," in J. Monod and J. Borek (eds.), *Microbes and Life,* pp. 50–55. New York: Columbia University Press.

Eldredge, N., and Gould, S. (1972). "Punctuated Equilibria: an Alternative to Phyletic/Gradualism," in T. Schopf (ed.), *Models in Paleobiology,* pp. 82–115. San Francisco: Freeman Cooper.

Furth, M. (1975). *Essence and Individual: Reconstruction of an Aristotelian Metaphysics,* chapter 11, duplicated for the meeting of the Society for Ancient Greek Philosophy, unpublished.

Ghiselin, M. (1966). "On Psychologism in the Logic of Taxonomic Controversies," *Systematic Zoology 15:* 207–15.

Ghiselin, M. (1969). *The Triumph of the Darwinian Method.* Berkeley: University of California Press.

Ghiselin, M. (1974). "A Radical Solution to the Species Problem," *Systematic Zoology 23:* 536–44.

Glass, B. (1959a). "Heredity and Variation in the Eighteenth Century Concept of the Species," in B. Glass, *et al.* (eds.), *Forerunners of Darwin,* pp. 144–72. Baltimore: The Johns Hopkins Press.

Glass, B. (1959b). "Maupertuis, Pioneer of Genetics and Evolution," in B. Glass, *et al.* (eds.), *Forerunners of Darwin,* 51–83. Baltimore: The Johns Hopkins Press.

Hacking, I. (1975). "The Autonomy of Statistical Law," talk delivered to The American Philosophical Association, Pacific Division, unpublished.

Hilts, V. (1973). "Statistics and Social Science," in R. Giere and R. Westfall (eds.), *Foundations of Scientific Method in the Nineteenth Century,* pp. 206–33. Bloomington: Indiana University Press.

Hull, D. (1965). "The Effect of Essentialism on Taxonomy: 2000 Years of Stasis," *British Journal for the Philosophy of Science 15:* 314–16; *16:* 1–18.

Hull, D. (1967). "The Metaphysics of Evolution," *British Journal for the History of Science 3, 12:* 309–37.

Hull, D. (1968). "The Conflict between Spontaneous Generation and Aristotle's Metaphysics." *Proceedings of the Seventh Inter-American Congress of Philosophy, 2* (1968): 245–50. Quebec City: Les Presses de l'Université Laval.

Hull, D. (1973). *Darwin and his Critics.* Cambridge, Mass.: Harvard University Press.

Hull, D. (1976). "Are Species Really Individuals?" *Systematic Zoology 25:* 174–91.

Hull, D. (1978). "A Matter of Individuality," *Philosophy of Science 45:* 335–60.

Ihde, A. (1964). *The Development of Modern Chemistry.* New York: Harper & Row.

Kripke, S. (1972). "Naming and Necessity," in D. Davidson and G. Harman (eds.), *Semantics of Natural Languages,* 253–355; 763–9. Dordrecht: Reidel.

Kripke, S. (1978). "Time and Identity." Lectures given at Cornell University, unpublished.

Lewontin, R. (1974). *The Genetic Basis of Evolutionary Change.* New York: Columbia University Press.

Lewontin, R. (1977). "Biological Determinism as a Social Weapon" in the Ann Arbor Science for the People Editorial Collective: *Biology as a Social Weapon,* pp. 6–20. Minneapolis, Minnesota: Burgess.

Lloyd, G. (1968). *Aristotle: The Growth and Structure of His Thought.* Cambridge: Cambridge University Press.

Lovejoy, A. (1936). *The Great Chain of Being.* Cambridge, Mass.: Harvard University Press.

Mayr, E. (1959). "Typological versus Population Thinking," in *Evolution and Anthropology: A Centennial Appraisal,* pp. 409–12. Washington DC: The Anthropological Society of Washington; also in Mayr (1976): 26–9 (text page references are to Mayr 1976).

Mayr, E. (1963). *Animal Species and Evolution.* Cambridge, Mass.: The Belknap Press of Harvard University Press.

Mayr, E. (1976). *Evolution and the Diversity of Life.* Cambridge, Mass.: Harvard University Press.

Meyer, A. (1939). *The Rise of Embryology.* Stanford, Calif.: Stanford University Press.

Minkowski, H. (1908). "Space and Time" in H. Lorentz, A. Einstein, *et al., The Principle of Relativity,* 73–91. New York: Dover.

Popper, K. (1972). *Objective Knowledge,* Oxford: Oxford University Press.

Preuss, A. (1975). *Science and Philosophy in Aristotle's Biological Works.* New York: Georg Olms.

Provine, W. (1971). *The Origins of Theoretical Population Genetics.* Chicago: University of Chicago Press.

Putnam, H. (1975). "The Meaning of 'Meaning'," *Mind, Language and Reality,* pp. 215–71. Cambridge: Cambridge University Press.

Quetelet, A. (1842). *A Treatise on Man and the Development of his Faculties.* Edinburgh.

Quine (1953a). "Identity, Ostension, Hypostasis" in *From a Logical Point of View,* pp. 65–79. New York: Harper Torchbooks.

Quine, W. (1953b). "Reference and Modality" in *From a Logical Point of View,* pp. 139–59. New York: Harper Torchbooks.

Quine, W. (1960). *Word and Object.* Cambridge, Mass.: MIT Press.

Rabel, G. (1939). "Long Before Darwin: Linne's Views on the Origin of Species," *Discovery, N.S., 2:* 121–75.

Ramsbottom, J. (1938). "Linnaeus and the Species Concept," *Proceedings of the Linnean Society of London,* 192–219.

Roger, J. (1963). *Les Sciences de la Vie dans la Pensée Française du XVIII Siècle.* Paris: Armand Colin.

Sober, E. (1982). "Realism and Independence," *Noûs 16:* 369–386.

Todhunter, I. (1865). *History of the Theory of Probability to the Time of Laplace.* New York: Chelsea Publishing.

Walker, H. (1929). *Studies in the History of Statistical Method.* Baltimore: Williams & Wilkins.

Wilson, E., and Bossert, W. (1971). *A Primer of Population Biology.* Sunderland, Mass.: Sinauer.

12

Temporally oriented laws

1. THE PROBLEM

Much has been written about laws that say that some quantity increases with time in the evolution of systems of a certain type. The second law of thermodynamics is the most famous example. Such laws are said to embed an asymmetry between earlier and later. In this paper, I want to discuss a rather different property of laws. Laws can allow us to calculate the future from the past or the past from the future.[1] Laws that do one of these but not the other I will call *temporally oriented.*

The strict second law is not temporally oriented. If I observe the present entropy of a closed system, I can infer that the entropy will be no less in the future *and* that it was no greater in the past. A law that posits a monotonic increase (or nondecrease) in a quantity permits inference in both directions.

Non-probabilistic laws that describe a conditional relationship between earlier and later are not temporally oriented. A law of the form 'If the system is in state E at an earlier time, then it will be in state L at a later time' supports inferences in both directions. A glimmer of the asymmetry we seek is to be found in the idea of conditional probability. A law that assigns a value to a probability of the form 'Pr(the system later is in state L/the system earlier is in state E)' permits forward-, but not backward-, directed inferences.[2] And, of course, precisely the opposite will be true if conditioning and conditioned propositions are reversed.[3]

In saying that 'Pr (the system later is in state L/the system earlier is in state E)' permits a forward, but not a backward, inference, I set to one side the possibility of using this law in a Bayesian format wherein

My thanks to Martin Barrett, James Crow, Carter Denniston, Ellery Eells, Malcolm Forster, and Richard Lewontin for useful comments.

prior probabilities plus the forward-directed law allow one to compute Pr(the system earlier is in state E/the system later is in state L). The point is that the forward-directed law, *by itself,* permits no such inference. Bayesians insist that priors are always available; their critics disagree. However, for the purposes of investigating the concept of a temporally oriented law, this controversy may be bracketed.

Are scientific laws often temporally oriented? Among laws that are temporally oriented, are more of them oriented to the future than to the past? When a law is temporally oriented, is this an artifact of our interests or does it reflect some objective feature of the world? These are the sorts of questions I wish to investigate.

2. A PROOF

Consider any three times t_1, t_2, and t_3, which occur in that order, and which are equally spaced. Let '$\Pr(t_i = x)$' denote the probability that the system at time t_i is in state x ($i = 1, 2, 3$; $x = 0, 1, 2, \ldots$). The laws[4] governing the evolution of the system will be given by conditional probabilities of the form $\Pr(t_i = x / t_j = y)$. I will say that the law is *forward-directed* if $i > j$ and *backward-directed* if $i < j$.[5]

A forward-directed law is *invariant under time translation* if $\Pr(t_2 = x / t_1 = y) = \Pr(t_3 = x / t_2 = y)$, for any equally spaced t_1, t_2, t_3 and for any x and y. That is, the 'date' at which a temporal interval begins or ends is irrelevant to calculating the probability of the system's state after some fixed amount of time has elapsed; all that matters is the system's (undated) initial state (specified in the conditioning proposition) and the amount of time that elapses between it and the end state.[6] Likewise, a backward-directed law is time translationally invariant if $\Pr(t_1 = x / t_2 = y) = \Pr(t_2 = x / t_3 = y)$, for any equally spaced t_1, t_2, t_3 and for any x and y.

It is worth noticing two features of this definition of time translational invariance. The definition requires that two conditional probabilities be equal, for any *fixed amount of spacing* between the three times and for any *states* that the system may occupy at a time. To see what these two constraints amount to, let us consider models in population genetics that relate the gene frequencies at one time in a population to some probability distribution of gene frequencies in the population some time later. An example is provided by models of neutral evolution, in which gene frequencies evolve by random walk (Crow and Kimura 1970; Kimura 1983). These models are time translationally invariant in the strong sense defined. First, the fixed amount of spacing between the three times may be any number of generations you please. Second, the model applies to all possible initial gene frequencies. That is, the theory of neutral evolution is time translationally invariant in the following sense:

> For any starting time t, for any number of generations n, and for any pair of gene frequency distributions i and j, Pr(Population is in state j at time $t + n$ / Population is in state i at t) = Pr(Population is in state j at time $t + 2n$ / Population is in state i at $t + n$).

This strong concept of time translational invariance will be assumed in what follows, although weaker notions of invariance are certainly possible. A law might be invariant for some amounts of spacing between the three events but not others; and it might be invariant for some possible states of the system but not others. A consequence of weakening the definition will be considered in due course.

I now will show that a system whose expected state[7] changes with time cannot have both a forward-directed time translationally invariant law and a backward-directed time translationally invariant law. If we assume that laws must be translationally invariant,[8] then this simplifies to: If a system's expected state changes, then it cannot have both a forward-directed probabilistic law and a backward-directed probabilistic law.

I'll begin by assuming that the system must be in one of two states (0 or 1) at any time; I'll dispense with this assumption shortly.

If the system possessed a backward-directed translationally invariant law, the following equality would have to hold:

(B) $$\Pr(t_1 = 1 / t_2 = 0) / \Pr(t_1 = 0 / t_2 = 0) = \Pr(t_2 = 1 / t_3 = 0) / \Pr(t_2 = 0 / t_3 = 0).$$

Bayes' theorem allows (B) to be expanded into the following:

$$\Pr(t_2 = 0 / t_1 = 1)\Pr(t_1 = 1) / \Pr(t_2 = 0 / t_1 = 0)\Pr(t_1 = 0) = \Pr(t_3 = 0 / t_2 = 1)\Pr(t_2 = 1) / \Pr(t_3 = 0 / t_2 = 0)\Pr(t_2 = 0).$$

If the system possessed a forward-directed time invariant law, it would be true that

(F) $$\Pr(t_2 = 0 / t_1 = 1) / \Pr(t_2 = 0 / t_1 = 0) = \Pr(t_3 = 0 / t_2 = 1) / \Pr(t_3 = 0 / t_2 = 0).$$

If we assume (F) and that the ratios mentioned in it are well defined but not zero, then the Bayesian expansion of (B) simplifies to:

(C) $$\Pr(t_1 = 1) / \Pr(t_1 = 0) = \Pr(t_2 = 1) / \Pr(t_2 = 0).$$

This last statement means that the expected state of the system does not change with time.

The proof easily generalizes to any finite number of states. Let 'x' and 'y' be variables that take states (0, 1, 2 . . .) as values. Parallel reasoning then entails that for any x and y:

$$\Pr(t_1 = x)/\Pr(t_1 = y) = \Pr(t_2 = x)/\Pr(t_2 = y).$$

If each pairwise ratio of this form must remain constant, then the expected state of the system cannot change.[9]

To say that the expected state does not change with time differs from saying that the state can be expected not to change. A particle doing a random walk on an open line has the same expected position throughout its history; on average it goes nowhere.[10] But this does not mean that we expect the particle to stay at the same location. Proposition (C) implies a lack of *directionality* in the system's probability trajectory.[11]

This result does not rule out evolution in the expected state of a system when the system has a forward-directed law *for one set of properties* and a backward-directed law *for some other set of properties.* Note that the argument considered forward- and backward-directed laws with respect to the same set of properties (t_i = 0, 1, 2 . . .). The point is that for each set of properties, a choice must be made between a forward-directed law and a backward-directed law, if laws are to be time translationally invariant and the system's expected state is to evolve.

The above argument cannot be carried through for systems all of whose transition probabilities are either 0 or 1; for in such cases the ratios in (F) either will equal zero or will not be well defined. The asymmetry we have established does not apply to *deterministic laws* (in at least one sense of that term).[12]

In the process here described, the effect of what happens at t_1 is itself the cause of what happens at t_3. But the idea that one and the same event is both effect and cause is not essential to the proof just offered. Consider, for example, the causal relationship between smoking (S) and dying of lung cancer (D). Smoking at one time causally contributes to dying sometimes later; but dying at one time does not contribute to subsequent smoking. Even so, it is easy to show that our proof applies. Two applications of Bayes' theorem yield the following equation:

$$\frac{\Pr(S/D)}{\Pr(-S/D)} = \left[\frac{\Pr(D/S)}{\Pr(D/-S)}\right]\left[\frac{\Pr(S)}{\Pr(-S)}\right]$$

The left side of the equation describes a ratio of backward-directed conditional probabilities. The first product term on the right describes a ratio of forward-directed conditional probabilities. Suppose this forward ratio is time translationally invariant. Then the ratio on the left side of the equation is time translationally invariant only if $\Pr(S)/\Pr(-S)$ is as well. I conclude that forward-directed and backward-directed conditional probabilities cannot both be time translationally invariant, if the probability of the cause (smoking, in this case) changes.

I mentioned earlier that my argument makes use of a rather strong concept of time translational invariance. It is worth considering the effect of weakening it in a certain way. Instead of considering time translational invariance as covering all amounts of spacing between the three times, let us consider the weaker idea of time translational invariance with respect to the temporal interval k (where k is a constant). If the forward- and backward-directed laws are both time translationally invariant with respect to k, then we can deduce that for any states x and y

$$\Pr(t_1 = x)/\Pr(t_1 = y) = \Pr(t_2 = x)\ /\Pr(t_2 = y),$$

where t_1 and t_2 are any two times separated by a temporal interval that is k units in length. In such a case, we cannot conclude that the system's expected state never changes, but only that it has the same value every k units of time. That is, with this weakened notion of invariance, we may conclude that the system's expected state is constant *or that it cycles with period k.* Not surprisingly, a weaker result is obtained for a weaker notion of invariance.

3. BREAKING THE SYMMETRY

It is fairly clear that science finds it natural to think in terms of forward-directed probabilities. We talk of the half-life of uranium. This is the amount of time it takes for the radioactivity of a sample to drop to half its initial value; it reflects a forward-directed probability that is assumed to be translationally invariant. In population genetics, Mendelism is a process by which parental genotypes confer probabilities on offspring genotypes; this, too, reflects a forward-directed probability assumed to be translationally invariant.

This temporal asymmetry in scientific concepts is also present in the concepts we use in everyday life. Roads are *dangerous;* skies are *threatening;* people are *friendly.* These familiar dispositional concepts, once they are understood probabilistically,[13] can be seen to describe forward-directed probabilities.

The proof of the previous section shows that we must choose between forward-directed and backward-directed laws. It appears that science and common sense have opted for the former. Is there some general feature of the world that explains why this should be so?

The probability of getting cancer if you smoke has changed through time, because other causal factors (for example, the amount of asbestos in the environment) have also changed. But once these other causal factors are held fixed, the conditional probability of getting cancer relative to them seems like it should be translationally invariant.[14]

The probability of having been a smoker, given that you have lung cancer, has also changed with time. But if we hold fixed the other effects of smoking (emphysema, for example), there is no reason to expect that the backward-directed conditional probability is temporally invariant.

One (rough) formulation of (forward-directed) *determinism* is the thesis that the state of the universe at time t_i is incompletely specified by a law, if the law assigns an intermediate value to a probability of the form $\Pr(t_j = x / t_i = y)$ (where $i < j$). The idea that completely specified causes generate time translationally invariant forward-directed laws makes no such deterministic assumption; the idea is that such probabilities should be stable under time translation, not that they should have values of 0 or 1.

Imagine an experimenter who runs a series of trials on an experimental setup one day, then waits a week and runs a second series of trials. Suppose that the two series of trials show markedly different frequency distributions. The experimenter reasons that although it is possible that the two series were generated by the same underlying probabilities, it strains one's credulity to think so. So the experimenter infers that the probabilities of outcomes have changed from one week to the next.

The natural inference to make here is *not* to attribute the change in probabilities to 'the passage of time' and let things go at that. Rather, what the experimenter will look for is a physical change in the experimental setup. Each series of trials is to be specified by its own conditional probability; the relevant conditioning propositions must differ. The problem is to find out how. We see here a practical consequence of accepting the idea that completely specified forward-directed probabilistic laws must be time translationally invariant.

The slogan 'same cause, same effect' is a common expression of the idea of determinism. I suggest that a vestige of this principle survives the dismantling of determinism; it is the idea of 'same cause, same probability of effect'. The slogan's converse – 'same effect, same cause' – was not much cited by those who accepted a deterministic world picture, though perhaps it should have been, since Newtonian physics is time symmetric. In any event, its probabilistic analog – 'same effect, same probability of cause' – is evidently not a plausible constraint on stochastic model building in the sciences.

So the preference for forward-directed laws reflects the assumption that 'same cause, same probability of effect' is true, but 'same effect, same probability of cause' is not. In each case, sameness of cause or of effect must be understood in terms of a complete specification; without this rider, neither principle is plausible.

I do not claim that the choice of forward-directed over backward-directed laws is explained by the acceptance of one of these principles but not the other. Rather, our preference for forward-directed laws and our opting for 'same cause, same probability of effect' are two expressions of the same underlying idea. Is there some basic feature of the world that explains both sorts of choices?

Let us look carefully at the example of Mendelism and describe with more care how probabilities are defined there. Consider a one-locus, two-allele example. There are three genotypes (AA, Aa, and aa) possible in each sex, so there are, in principle, nine possible parental pairs; if we ignore the sex of each parent and consider only the genotypes they contribute, then there are six pairs. For each pair we can say how probable it is that an offspring should have a particular genotype. The rows and columns denote the parental genotypes and the cells provide the probabilities of offspring genotypes, conditional on their coming from a particular cross:[15]

		Female		
		AA	Aa	aa
Male	AA	1(AA)	1/2(AA); 1/2(Aa)	1(Aa)
	Aa	—	1/4(AA); 1/2(Aa); 1/4(aa)	1/2(Aa); 1/2(aa)
	aa	—	—	1(aa)

Notice that the Mendelian process does *not* allow you to compute probabilities for parental genotypes from the genotypes of offspring. If the sons and daughters are heterozygotes, how probable is it that the parents were also heterozygotes? This question is unanswerable, until prior probabilities are given for the genotypes of parental pairs.[16] The asymmetry under consideration can be illustrated as follows:

Genotype of parental pair + Mendelism ⟶ Probability of Offspring Genotype

Probability of Parental Genotypes ⟵/⟶ Mendelism + Genotype of offspring

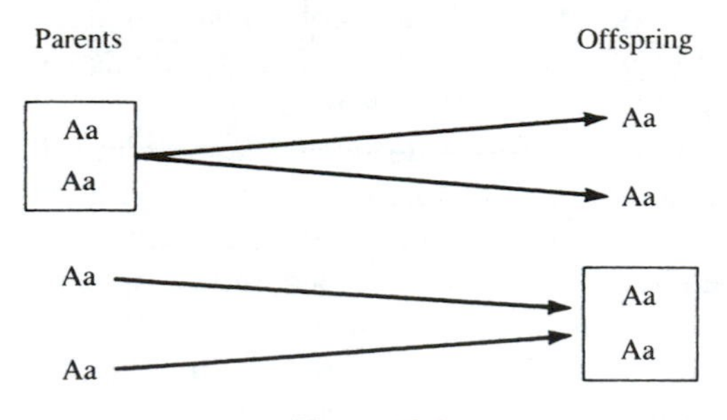

Figure 12.1

As noted before, an assumption of completeness is needed here. If there is unequal mortality among offspring genotypes and if the offspring genotypes are censused sometime after the egg stage, then the predicted Mendelian ratios for the offspring will be violated. The assumption, therefore, is that no such interferences impinge.

Let us now look at this asymmetry from a different angle. Suppose that two heterozygote parents produce two heterozygote offspring. We can think of the parental pair as the common cause of two effects, or we can think of the offspring pair as the common effect of two causes (Figure 12.1). The genotype of the parental pair not only confers a probability on the offspring genotypes; in addition, the genotype of the parental pair renders the genotypes of the two offspring (O_1 and O_2) independent of each other:

$$\begin{aligned}&\Pr(O_1 \text{ is Aa } \& \ O_2 \text{ is Aa / parents are Aa})\\&\quad = \Pr(O_1 \text{ is Aa / parents are Aa})\Pr(O_2 \text{ is Aa / parents are Aa}).\end{aligned}$$

That is, the common cause forms a *conjunctive fork* with its two effects (Reichenbach 1956; see Essay 8 for further discussion).

In contrast, the two offspring genotypes do *not* render the genotypes of the two parents (P_1 and P_2) statistically independent of each other:

$$\begin{aligned}&\Pr(P_1 \text{ is Aa } \& \ P_2 \text{ is Aa / offspring are Aa})\\&\quad \neq \Pr(P_1 \text{ is Aa / offspring are Aa})\Pr(P_2 \text{ is Aa / offspring are Aa}).\end{aligned}$$

I believe that this Mendelian asymmetry is typical of a larger pattern: *A common cause often forms a conjunctive fork with its joint effects, but a common effect rarely forms a conjunctive fork with its joint causes* (Sober and Barrett 1992). In the next section, I'll defend this pair of claims. For the present, I want to show how this asymmetry, if true, explains why temporally oriented laws are usually forward-directed.

If I want to describe the probability of an offspring's genotype, given the genotype of its parents, *I do not have to take into account what else is true simultaneously with that offspring.* I do not need to know what that offspring's sibs are like; I also do not need to know the prior probability of offspring genotypes. This is a consequence of the fact that the first fork displayed above is conjunctive.

In contrast, if I wish to describe the probability of a parent's genotype, given the genotypes of its offspring, I must take into account facts that pertain to the parental generation. I need to know the pattern of mating (random or assortative) and the prior probability of parental genotypes. This is a consequence of the fact that the second fork is not conjunctive; the genotypes of the offspring do not render statistically irrelevant all other facts simultaneous with the parent's having the genotype it does.

Because the second fork is not conjunctive, the probability of a cause, given its effects, cannot be purely a matter of law. The happenstance of what else is true simultaneously with the cause must also be taken into account. In contrast, the probability of an effect, given its causes, may be a matter of law, since the happenstance of what else is true simultaneously with the effect may be ignored, if the fork is a conjunctive one. My suggestion is that laws are temporally oriented towards the future because of this asymmetry between forks.

In the example just discussed, I have treated the rules of Mendelism as providing a law governing the reproductive behavior of the organisms in a population. There is room to quarrel with the idea that organisms obey Mendelian rules purely as a matter of law. After all, the population is the result of evolution and it is perfectly reasonable to see its conformity to Mendelian patterns as a consequence of that evolution. This makes Mendelism a matter of happenstance, not a matter of law (Beatty 1981).

Although this raises interesting questions about the role of laws in evolutionary theory, it does not undermine the main point I wish to argue for here. In a Mendelian system, one can describe a forward-directed conditional probability that does not depend for its correctness on the prior distributions of genotypes at either time or on the system of mating; this cannot be done for the backward-directed conditional probability. If the prior distributions (or the system of mating) are not matters of law, then forward-directed conditional probabilities may be laws, but backward-directed conditional probabilities cannot be.

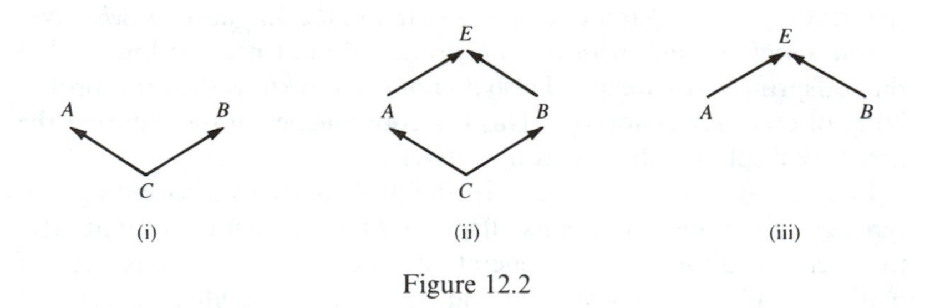

Figure 12.2

4. FORKS

Reichenbach (1956) advanced two theses that together entail an asymmetry between future and past. First, he maintained that if A and B are simultaneous and correlated events, they must have a common cause that renders them conditionally independent of each other. Reichenbach's second claim is that if A and B have a common effect, this common effect will 'usually' form a conjunctive fork with A and B. What is not allowed is that A and B lack a common cause but form a conjunctive fork with a common effect. That is, Reichenbach maintained that of the three possible patterns that conjunctive forks might display, the first two are common, whereas the third is impossible (Figure 12.2). As Reichenbach put it, (conjunctive) forks open to the future are possible, whereas forks open to the past are not.

Reichenbach's first claim, which he called the *principle of the common cause,* is incompatible with results stemming from quantum mechanics (van Fraassen 1982). In addition, it conflicts with a perfectly 'classical' probabilistic phenomenon: Two causally independent processes may be correlated because each shows a monotonic increase in some quantity (see Essay 8). Yet consistent with these corrections, a rather Reichenbachian thesis can be advanced concerning common causes and their joint effects: Quantum mechanical phenomena aside, a complete specification of the causes must render the effects conditionally independent of each other.

Reichenbach's second claim is also off the mark. Conjunctive forks open to the past are not impossible. Even so, joint causes (whether they share a common cause) *only rarely* form conjunctive forks with their common effect (Sober and Barrett 1992).

To see why, let us schematize the problem as follows. A and B are joint causes of E. First, I define the four conditional probabilities of the form $\Pr(E / \pm A \& \pm B)$ as follows:

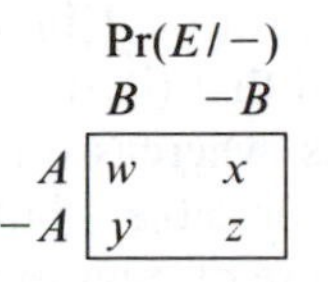

$\Pr(E/-)$

	B	$-B$
A	w	x
$-A$	y	z

I'll call these probabilities the *efficacies of the causes.* They are all forward-directed (assuming that cause precedes effect); no assumption will be made as to whether they are invariant over time. The four probabilities of the form $\Pr(\pm A\&\pm B)$ I'll call the probabilities of *combinations of causes.*

AEB is a conjunctive fork iff the following equality is true:

$$\Pr(A\&B/E) = \Pr(A/E)\Pr(B/E).$$ [17]

Bayes' theorem allows these three conditional probabilities to be expressed as follows:

$$\begin{aligned}
\Pr(A\&B/E) &= \Pr(E/A\&B)\Pr(A\&B)/\Pr(E) \\
&= w\Pr(A\&B)/\Pr(E) \\
\Pr(A/E) &= \Pr(E/A)\Pr(A)/\Pr(E) \\
&= [\Pr(E/A\&B)\Pr(B/A) \\
&\quad + \Pr(E/A\&-B)\Pr(-B/A)]\Pr(A)/\Pr(E) \\
&= [w\Pr(B/A) + x\Pr(-B/A)]\Pr(A)/\Pr(E) \\
\Pr(B/E) &= \Pr(E/B)\Pr(B)/\Pr(E) \\
&= [\Pr(E/A\&B)\Pr(A/B) \\
&\quad + \Pr(E/-A\&B)\Pr(-A/B)]\Pr(B)/\Pr(E) \\
&= [w\Pr(A/B) + y\Pr(-A/B)]\Pr(B)/\Pr(E)
\end{aligned}$$

So the fork is conjunctive precisely when

$$\begin{aligned}
w\Pr(A\&B) &= [w\Pr(A\&B) + x\Pr(A\&-B)][w\Pr(A\&B) \\
&\quad + y\Pr(-A\&B)]/\Pr(E).
\end{aligned}$$

This simplifies to

$$wz/xy = \Pr(A\&-B)\Pr(-A\&B)/\Pr(A\&B)\Pr(-A\&-B). \qquad (*)$$

Note that the left-hand side of (*) describes a relationship among the *efficacies* of the causes, while the right-hand side describes a relationship among the *frequencies* of the causes.

The frequencies of causes often undergo change; so do the efficacies of causes (at least when the causes are incompletely specified). My claim is that when the frequencies and/or efficacies of causes change, the changes are almost never coordinated so as to keep (*) true. If *AEB* is a conjunctive fork at a given time, this is a mathematical accident that is soon canceled by changes in the system at hand.

So even though Reichenbach's conjectured fork asymmetry is not correct, another can be substituted in its stead (Sober and Barrett 1992). Reichenbach claimed that (i) and (ii) are common patterns displayed by conjunctive forks, whereas (iii) is impossible. I have argued that (ii) and (iii) are both possible – but are very rare – and that (i) requires hedging only because of quantum mechanical phenomena.

I now want to apply this general schema to the Mendelian case discussed before. Reichenbach's exposition and the result just described assume that the events in the fork are dichotomous. But in the simplest of Mendelian cases, this isn't so. The parental pair has six possible states and each offspring has three. Yet it is clear enough how to generalize Reichenbach's idea to cover this situation.

I'll begin with the forward fork formed by a common cause and its two effects. The two parents are heterozygotes and we wish to describe the probabilistic relationship between them and their two offspring, whom we'll assume are heterozygotes as well. The parental pair is the cause and the two offspring are the effects.

The conjunctive fork idea requires that the different possible states that the parental pair might occupy each render the offspring's genotypes independent of each other. Focusing on the case in which both parents are heterozygotes, this means that

$$\text{(FOR)}\ \Pr(\mathrm{Off}_1 = \mathrm{Aa}\ \&\ \mathrm{Off}_2 = \mathrm{Aa} / \mathrm{Par} = \mathrm{Aa\&Aa}) = \Pr(\mathrm{Off}_1 = \mathrm{Aa} / \mathrm{Par} = \mathrm{Aa\&Aa})\Pr(\mathrm{Off}_2 = \mathrm{Aa} / \mathrm{Par} = \mathrm{Aa\&Aa}).$$

It is quite clear that (FOR) is a standard assumption in models of the Mendelian process.

Let us turn now to the backward fork, formed by a common effect and its two causes. We wish to determine under what circumstances the genotypes of the offspring form a conjunctive fork with the genotypes of the two parents (P_1 and P_2). For simplicity (but without loss of generality), I'll focus on the case of a single heterozygous offspring whose two parents are heterozygotes as well.

If the fork is to be conjunctive, the different possible states of the offspring must each render the parental genotypes independent of each other. That is, the fork will be conjunctive only if

$$\text{(BACK)}\quad \Pr(P_1 = \mathrm{Aa}\ \&\ P_2 = \mathrm{Aa} / \mathrm{Off} = \mathrm{Aa}) = \Pr(P_1 = \mathrm{Aa} / \mathrm{Off} = \mathrm{Aa})\Pr(P_2 = \mathrm{Aa} / \mathrm{Off} = \mathrm{Aa}).$$

I prove in the Appendix that (BACK) is true if and only if:

$$\Pr(P_2 = \mathrm{Aa} / P_1 = \mathrm{Aa}) / \Pr(P_2 = \mathrm{Aa}) = 1 / [2\Pr(\mathrm{Off} = \mathrm{Aa})].\quad (**)$$

The left side of (**) describes the degree of assortative mating that occurs among parents; it tells you how much the choice of mates departs from randomness. The right side describes the frequency of heterozygotes among the offspring. Notice that for any degree of assortative mating, there is exactly one point-value for the offspring frequency of heterozygotes that satisfies the requirements for a conjunctive fork.

If like always mates with like, then the assortative mating is said to be perfect and heterozygosity declines to zero. If the positive assortative mating falls short of this absolute degree, then heterozygosity declines until it reaches an equilibrium value (Crow and Kimura 1970, p. 144). And of course other evolutionary forces besides pattern of mating can modify the frequency of heterozygotes.

If the assortative mating is perfect, (**) reduces to

$$2 = \Pr(P_2 = Aa) / \Pr(Off = Aa)$$

and indeed it is a property of this system of mating that heterozygosity is halved in every generation. So the requirement provided by (**) is satisfied in this case. However, the conjunctive fork idea imposes other demands. In particular, the other possible offspring genotypes must render the parental genotypes independent of each other:

$$\Pr(P_1 = Aa \;\&\; P_2 = Aa / Off = AA)$$
$$= \Pr(P_1 = Aa / Off = AA)\Pr(P_2 = Aa / Off = AA).$$

This will not be true when there is perfect positive assortative mating.

It is not impossible for a population experiencing positive (though imperfect) assortative mating to satisfy (**) over the long haul. What is required is that the intensity of positive association between mates should adjust itself so as to satisfy a criterion specified in terms of the offspring frequency of heterozygotes. The rarer heterozygotes become, the more monomaniacally must like mate with like (all this in accordance with an exact *quantitative* formula). Rube Goldberg devices for achieving this coordination are not ruled out *a priori,* but it is not surprising that no living system happens to possess one.[18]

The above argument about the backward fork focuses on the case in which two parents have a single offspring. What is the effect of increasing the number of offspring? If the parental pair has an infinite number of offspring, the likelihood terms (the probability of offspring genotypes conditional on the genotypes of the parents) approach 1 or 0, and the fork becomes conjunctive degeneratively. But short of this limiting case, the conclusion that conjunctive forks are hard to come by remains in place.

Correlation between sibs is the inevitable outcome of Mendelian reproduction (in a segregating population). Correlation between parents is not inevitable, but is the familiar pattern that is called assortative mating. When sib genotypes are correlated, it is easy to explain why this is so by invoking a model that says that each parental pair forms a conjunctive fork with the children it produces. A model that says that a child forms a conjunctive fork with its two parents allows one to deduce the pattern of association found between the parents. One reason for claiming that this model is not an explanation is to insist that later events don't explain events that happen earlier. But, in addition, it is worth noting that the purely probabilistic relationships invoked in such a model are rarely if ever satisfied in nature.

5. CONCLUDING REMARKS

The concept of *law* naturally carries with it the idea of time translational invariance. This requirement, by itself, induces no temporal orientation in the laws we posit. The same can be said of the idea of *probability*. That a law should be stochastic, and that it should say of certain systems that their expected state changes with time, also induces no temporal asymmetry. But these two ideas, each of them innocent when taken singly, require that temporal symmetry be broken when they are taken together.

It is a separate question which way the symmetry is broken. If probabilistic laws were backward-directed about as often as they are forward-directed, we might not expect there to be a single underlying explanation for why symmetry is broken in the way it is. But if most (or all) temporally oriented laws are forward-directed, the suspicion arises that there is a single explanation of this fact. I have floated the idea that the source of bias in favor of forward-directed laws is to be found in an asymmetry concerning causality. We can talk of the probability of an effect, given a cause, and also of the probability of a cause, given an effect. With incomplete descriptions of both cause and effect, neither of these conditional probabilities can be expected to be time translationally invariant. However, when the causal facts are completely circumscribed, it is plausible to maintain that the forward-directed conditional probability is time translationally invariant. The same cannot be said for the backward-directed probability when the description of the effects is rendered complete. The reason for this, I suggest, is that (completely specified) common causes often form conjunctive forks with their joint effects, but (completely specified) common effects rarely form conjunctive forks with their joint causes. This may go some way to explaining why temporally oriented laws are forward-directed.

At the beginning of this essay, I raised the question of whether the temporal asymmetry described here is an artifact of our interests or an objective feature of the world. Different aspects of my argument answer this question in different ways. The proof with which I began owes nothing to our interests; it is a mathematical fact that systems of the kind specified cannot have both forward-oriented and backward-oriented laws, if laws must be time translationally invariant. I then argued that science opts for forward-directed laws because conjunctive forks open to the future are common while conjuctive forks open to the past are rare. This difference between forks derives, in part, from the fact that the efficacies of causes are not tightly coupled with their frequencies, and again I take this relative independence to be an objective feature of the world. But what of our choice of *descriptors?* What entitles us to carve up the world into efficacies and frequencies that are not tightly coupled? Could not a 'gruification' of familiar terminology produce a vocabulary in which efficacies and frequencies are bound together, thus ensuring that conjunctive forks open to the past are as common as dirt? Having pursued the question of objectivity back this far, I will not attempt to pursue it farther. For now, my conclusion is a conditional one: *Given* the descriptors we use of the probabilities and efficacies of causes, it is an objective feature of the world that temporally oriented laws are oriented towards the future.

APPENDIX

I want to prove that

$$\text{(BACK)} \quad \Pr(P_1 = Aa \;\&\; P_2 = Aa \,/\, Off = Aa) = \Pr(P_1 = Aa \,/\, Off = Aa)\Pr(P_2 = Aa \,/\, Off = Aa)$$

is true if and only if

$$\Pr(P_2 = Aa \,/\, P_1 = Aa) \,/\, \Pr(P_2 = Aa) = 1 \,/\, [2\Pr(Off = Aa)]. \quad (**)$$

Bayes' theorem allows (BACK) to be rewritten as:

$$\tfrac{1}{2}\,[\Pr(P_1 = Aa \;\&\; P_2 = Aa)] \,/\, \Pr(Off = Aa) = [\Pr(Off = Aa \,/\, P_1 = Aa)\Pr(P_1 = Aa) \,/\, \Pr(Off = Aa)]^2.$$

Note that

$$
\begin{aligned}
&\Pr(\text{Off} = \text{Aa} / \text{P}_1 = \text{Aa}) \\
&= \Pr(\text{Off} = \text{Aa} / \text{P}_2 = \text{AA} \;\&\; \text{P}_1 = \text{Aa})\Pr(\text{P}_2 = \text{AA} / \text{P}_1 = \text{Aa}) \\
&\quad + \Pr(\text{Off} = \text{Aa} / \text{P}_2 = \text{Aa} \;\&\; \text{P}_1 = \text{Aa})\Pr(\text{P}_2 = \text{Aa} / \text{P}_1 = \text{Aa}) \\
&\quad + \Pr(\text{Off} = \text{Aa} / \text{P}_2 = \text{aa} \;\&\; \text{P}_1 = \text{Aa})\Pr(\text{P}_2 = \text{aa} / \text{P}_1 = \text{Aa}) = \tfrac{1}{2}.
\end{aligned}
$$

This allows (BACK) to be further simplified to

$$
\tfrac{1}{2}\,[\Pr(\text{P}_1 = \text{Aa} \;\&\; \text{P}_2 = \text{Aa})\Pr(\text{Off} = \text{Aa})] / [\Pr(\text{P}_1 = \text{Aa})\Pr(\text{P}_1 = \text{Aa})] = \tfrac{1}{4},
$$

from which (**) follows.

NOTES

1. Or neither. But I'll ignore laws of simultaneous compossibility.
2. One can infer the later state of the system using this conditional probability if the conditional probability is high or low, once one has observed that the conditioning proposition is true. And even if the probability is middling, the conditional probability plus the observation allow one to assign a probability to the conditioned proposition.
3. The idea of a law's subserving one sort of inference but not the other will be clarified presently.
4. Nothing very demanding is implied by talk of 'laws' here. Issues about the modal status of laws don't matter, for example.
5. A forward-directed law, thus defined, can permit a backward inference based on likelihood, not probability. Suppose one observes that the system is presently in state *s;* this supports the hypothesis that it was earlier in state *i* better than it supports the hypothesis that it was earlier in state *j,* if Pr(system is now in state *s* / system was earlier in state *i*) > Pr(system is now in state *s* / system was earlier in state *j*).
6. In the theory of Markov processes, such laws are said to be *time stationary.*
7. By 'expected state', I mean the state's *mathematical expectation.* In general, the mathematical expectation of a quantity is not the value we would expect the quantity to have. Consider a population in which half the parents have one child and half have two. The expected number of offspring of a parent drawn at random from this population is 1.5, although we would not expect any parent to have that number of children.
8. Earman (1986) describes a number of writers who hold to this principle. He does not dissent from it. If laws describe causal variables, then the principle embodies the idea that the date of an event is not a causal vari-

able. I do not claim that this idea is an *a priori* constraint on our concept of law, plausible though it is in scientific practice.

9. Kemeny and Snell (1960, pp. 26, 105) point out that if a forward process is a Markov chain, the backward process will not generally be a Markov chain. They note that symmetry is restored if the process begins at equilibrium.
10. See Berg (1983) for details. The same point holds for random genetic drift. The initial gene frequency is the expected value throughout the population's evolution; yet the probability of the gene's going to an absorbing state (0% or 100%) increases with time. See Crow and Kimura (1970).
11. To see what is involved in this claim, consider a random walk on an open line, with the marker beginning at position 0. After a unit of time elapses, there is a chance of moving one unit to the left and an equal chance of moving one unit to the right. The process may continue for as long as you please. What is the expected location of the marker on the line from $+\infty$ to $-\infty$ after n units of time elapse? The expected location is at location 0. The longer the process continues, the less probable it is that the marker is at location 0. Yet this remains its expected position.

    Suppose the first step of this process, during one of its realizations, has the marker move one step to the left. Its expected position thereafter, conditional on this fact, is $+1$. But earlier, its expected position was 0. Isn't this a case in which the expected position has changed?

    No – not in the sense intended. Conditional on the position of the marker at any time (for example, the beginning of the process), we can ask what the expected position is after n moves. The point is that the answer to this question does not depend on n. It is in this sense that the expected position is the same throughout the process.
12. In the language of the theory of Markov processes, the above result applies to systems that are *ergodic:* for each pair of states i and j that the system might occupy, there is a nonzero probability that a system beginning in state i will remain there and a nonzero probability that it will go into state j. I believe that the result also applies to Markov chains that are nonergodic, but *decomposable;* this is the case in which the state space can be divided into regions, between which there is no chance of passing, but within which each state is accessible to every other.
13. I suggest that dispositional concepts like *solubility* should be understood probabilistically. If indeterminism were to imply that a lump of sugar immersed in water under normal conditions has a tiny but nonzero probability of *not* dissolving, I don't think we would conclude that sugar is not water soluble (Sober 1984).
14. See Eells (1991) for discussion of this point.
15. I am here ignoring the phenomenon of meiotic drive; that would complicate the example, but would not affect the main point.

16. The probability of a parental pair is itself a result of two factors: the probabilities of genotypes within each sex and the system of mating.
17. The conjunctive fork idea also requires that $\Pr(A\&B/-E) = \Pr(A/-E)\Pr(B/-E)$, but the argument would not be affected by considering this condition separately.
18. Although there is dispute among population geneticists about how assortative mating ought to be represented mathematically, one standard model postulates that a fraction r of each genotype pairs up with the like individuals and the remainder $(1 - r)$ mates at random (Crow and Kimura 1970, p. 144); r is here said to be the coefficient of assortative mating. Under this arrangement

$$\Pr(P_1 = Aa \;\&\; P_2 = Aa) = r\Pr(P_1 = Aa) + (1 - r)\Pr(P_1 = Aa)^2.$$

If we let $p = \Pr(P_1 = Aa)$ be the population frequency of Aa in the current generation and p' be the frequency in the next generation, then the condition for a conjunctive fork (**) becomes

$$r + (1 - r)p = p/2p'.$$

The frequency of heterozygosity in the next generation can be expressed as a function of the heterozygosity in the previous generation

$$p' = rp/2 + (1 - r)p_0,$$

where p_0 is the frequency of heterozygosity at the beginning of the process. This allows (**) to be restated as:

$$r + (1 - r)p = p/2p' = p/[rp + 2(1 - r)p_0].$$

As in the simpler treatment provided in the text, it is extremely improbable that an evolving population should satisfy this requirement.

REFERENCES

Beatty, J. (1981). 'What's Wrong with the Received View of Evolutionary Theory?', *PSA 1980,* Vol. 2. Philosophy of Science Association, East Lansing, MI, pp. 397–426.

Berg, H. (1983). *Random Walks in Biology.* Princeton: Princeton University Press.

Crow, J., and M. Kimura (1970). *Introduction to Population Genetics Theory,* Minneapolis: Burgess.

Earman, J. (1986). *A Primer on Determinism.* Dordrecht: Reidel.

Eells, E. (1991). *Probabilistic Causality.* Cambridge: Cambridge University Press.

Kemeny, J., and J. Snell (1960). *Finite Markov Processes,* Princeton: Van Nostrand.

Kimura, M. (1983). *The Neutral Theory of Molecular Evolution.* Cambridge: Cambridge University Press.

Reichenbach, H. (1956). *The Direction of Time.* Los Angeles and Berkeley: University of California Press.
Sober, E. (1984). *The Nature of Selection: Evolutionary Theory in Philosophical Focus.* Cambridge: MIT Press.
Sober, E., and M. Barrett (1992). 'Conjunctive Forks and Temporally Asymmetric Inference', *Australasian Journal of Philosophy, 70,* 1–23.
Van Fraassen, B. (1982). 'The Charybdis of Realism: Epistemological Implications of Bell's Inequality', *Synthese 52,* 25–38.

Index

Index

Index